Liquid Crystals and Display Technology

Edited by Morteza Sasani Ghamsari and Irina Carlescu

Published in London, United Kingdom

IntechOpen

Supporting open minds since 2005

Liquid Crystals and Display Technology
http://dx.doi.org/10.5772/intechopen.77795
Edited by Morteza Sasani Ghamsari and Irina Carlescu

Contributors
Guanming Yuan, Zhengwei Cui, Nikolaos Papadopoulos, Kris Myny, Pawel Malinowski, Lynn Verschueren, Tung Huei Ke, Auke Kronemeijer, Jan Genoe, Wim Dehaene, Rajendrasing Rajesing Deshmukh, Anuja Katariya-Jain, Byoungchoo Park, W. S. Lee, Seo Yeong Na, Jaewoo Park, In-Gon Bae, Mercedes Pérez Méndez, José Fayos Alcañiz, Chi Ming Che, Huiyang Li, Lei Dai, Byoungki Choi, Yoonhyun Kwak, Yong-Suk Cho, Liangliang Yan, Tsz-Lung Lam, Irina Carlescu

First published in London, United Kingdom, 2020 by IntechOpen
IntechOpen is the global imprint of INTECHOPEN LIMITED, registered in England and Wales, registration number: 11086078, 5 Princes Gate Court, London, SW7 2QJ, United Kingdom
Printed in Croatia

British Library Cataloguing-in-Publication Data
A catalogue record for this book is available from the British Library

Additional hard and PDF copies can be obtained from orders@intechopen.com

Liquid Crystals and Display Technology
Edited by Morteza Sasani Ghamsari and Irina Carlescu
p. cm.
Print ISBN 978-1-78985-367-4
Online ISBN 978-1-78985-368-1
eBook (PDF) ISBN 978-1-78985-505-0

For EU product safety concerns:
IN TECH d.o.o., Prolaz Marije Krucifikse Kozulić 3, 51000 Rijeka, Croatia, info@intechopen.com or visit our website at intechopen.com.

Meet the editors

Dr. Morteza Sasani Ghamsari is a senior researcher in the Photonics and Quantum Technologies Research School of Iranian Nuclear Science and Technology Research Institute. His research focuses on photonic materials including metamaterials, quantum dots, and plasmonic nanomaterials that can be used in a wide range of nanophotonics applications. His recent interests also include nano-bioimaging, 3D printing, nanostructures for tissue engineering (ZnO, TiO_2, etc.) and biomaterials including carbon, graphene, and diamond quantum dots. He is an editorial board member and reviewer for different international journals and has collaborated with local and international academics/researchers on post-graduate research projects. He has edited four books and published four chapters and more than 105 articles in scientific journals and reviewed conference proceedings. His papers have been cited more than 2100 times with h-index 23 and i-10 index 41 (Google Scholar).

Dr. Irina Cârlescu received an MS in Ecological Catalysis in 2000 and a PhD in Organic Chemistry in 2005. Her areas of research interest include synthesis and characterization of liquid crystals with applications in opto-electronics, azobenzene compounds, glycoconjugates and ferrocene derivatives. She is a reviewer for Liquid Crystals, *Crystals, Molecules and Medicinal Chemistry Research*. She is also co-author of twenty-seven articles in peer-reviewed journals, h-index = 8.

Contents

Preface

As output devices, displays are usually used to convey originated information from a machine to a person. The ability of display to provide valuable and reliable information in real time is very important for better tracking of information and facilitating rapid decision-making. Displays are generally used in different devices such as televisions, laptops, monitors, mobiles, and smartphones. From 1922 to now, the display industry has experienced rapid growth and expansion with no signs of slowing down. For example, plasma displays, field emission displays, and electronic papers are examples of the continued expansion of display technology. In addition, organic light-emitting diodes (LEDs) and Digital Light Processing (DLP) technology are other contenders for leadership in display technology. Many believe that the microLED is the next generation of display technology, and leading players in LED, display, original equipment material (OEM), and others are pursuing it. Differentiated from liquid crystal display (LCD) and organic light-emitting diode (OLED), microLED is considered to be the only display technology that has no size limitation. Nevertheless, new trends such as LCD displays have increased demand for displays due to their light weight, low operating power, and compact design. LCD displays are applied in almost all displays smaller than 65" and therefore they dominate the current display market. However, they sufer from their intrinsic limitation to go larger in size. Any small-screened electronics such as digital watches, cell phones, and laptops can be produced by LCDs. The liquid crystals are a different state of matter and exhibit different molecular arrangements from the liquid and solid states. Liquid crystals promise to fill a vast array of uses in the future. Due to the progressive physical, chemical, electrical, and optical properties of liquid crystals, they exhibit various electronic and optical functions that make them candidates for a variety of engineering applications such as displays, sensors, energy-related technologies, electro-photonics, and so on. They can be used to produce faster-responding displays and serve as building blocks of next-generation optoelectronic devices. For instance, QD-LCDs are employed in TVs, and an increasing number of consumers prefer to buy this kind of TV. Other kinds of displays have different advantages and disadvantages. High-resolution images can be produced by field emission displays without a bulky appearance. Excellent quality images can be generated by plasma displays on very large screens. LED displays are already used in huge public displays. To compete with them, microLEDs should identify their unique value propositions or value proposition combinations to offer advantages. In addition, it seems that OLEDs are taking an increasing market share mainly in smart phone displays. In our modern world, touchscreen displays are used in a wide variety of devices and are standard on handheld gadgets like smartphones and tablets. Accordingly, high-quality displays can drive technology evolution in which new approaches and innovative ideas in information presentation will be realized.

This book, *Liquid Crystals and Display Technology*, provides an overview of the recent advances in the synthesis of liquid crystals and displays manufacturing and their emerging applications for output devices. This book has two sections. In the first, the chapters focus on nematic liquid crystals. The first chapter discusses fundamental (and already well-known) issues regarding uniaxial and biaxial nematic phases. The next three chapters examine alignment and controlling of nematic microdoplets dispersed in a polymer matrix, chemical modifications of helical cholesteric liquid crystalline polyesters, and formation and characterization of polycyclic aromatic hydrocarbons-based mesophase pitch with optical anisotropy.

The second section contains three chapters. The first chapter in this section considers the additional functionalities beyond the regular display functions of an active matrix organic light-emitting diode (AMOLED) display. The second chapter describes the use of a nonporous, homogeneous, smooth, and easily processable graphene layer as a source contact together with an emissive channel layer of vertical-type OLETs (VOLETs). With a functionalized graphene source, the chapter shows that the full-surface electroluminescent emission of a VOLET can be effectively controlled by gate voltage with a high luminance on/off ratio. Finally, the last chapter provides an overview of tetradentate platinum(II) emitters as a promising class of metal-organic phosphorescent dopants for OLEDs. This book presents the most attractive and versatile technological developments in the field of liquid crystals and display technology to provide a better understanding of the currently ongoing research in related fields.

Morteza Sasani Ghamsari
Photonics and Quantum Technologies Research School,
Nuclear Science and Technology Research Institute,
Iran

Irina Carlescu
Gheorghe Asachi Technical University of Iași,
Romania

Section 1

Liquid Crystals

Chapter 1

Introductory Chapter: Nematic Liquid Crystals

Irina Carlescu

1. Nematic phase

The nematic (N) liquid crystalline phase is technologically the most important of the well-known and widely studied mesophases (nematic, smectic, cholesteric, and columnar). Nematics are also the most used, because they illustrate the best dual nature of liquid crystals. Hence, the molecules that form LC mesophases (intermediate states between the crystalline and liquid phase) display a unique combination of properties between long-range order and mobility, the basis of the numerous technical applications [1]. However, this combination constitutes as well an essential requirement for living matter, considering liquid crystals play a significant role in biomolecule's assembling, e.g., smectic phases (in phospholipid bilayer in the cell, protein filament), columnar phases (in DNA), or nematic phases (in chitin, collagen, cellulose, viruses, and silk) [2, 3].

Moreover, the ability of liquid crystal molecules to respond under weak external stimuli (temperature, electric or molecular adsorbates) stimulated the intellectual collaboration between specialists as chemists, physicists, or electrical engineers. As a result of joining of liquid crystal science and other fields, new opportunities for applications have been developed, involving polymers, colloids, or surfactants [4]. Hence, soft responsive materials based on surface-induced ordering transitions using nematic liquid crystal droplets dispersed within a medium or encapsulated into polymeric shells have been a promising perspective for experimental research [5].

The formation of liquid crystalline phase or mesomorphism implies the transition of a pure compound from an ordered crystalline state to a disordered liquid in two events: by a change in temperature and melting, the case of thermotropic liquid crystals, and by adding a suitable solvent to a mesogen and dissolving, the case of lyotropic liquid crystals. Thermotropic liquid crystalline state occurs mostly in compounds with pronounced molecular anisotropy. Lyotropic systems, representing two- or multi-component, occur when dissolved mesogenic amphiphiles self-assemble into ordered micelles.

Because of high mobility, nematic phases show a low viscosity, very similar to isotropic liquids, with the difference that the parallelism of the long axes induces the anisotropy of many physical properties. Hence, the nematic liquid crystals are anisotropic in respect to optical properties (double refraction), viscosity, magnetic and electric susceptibility, and electric and thermal conductivities [6].

The nematic phase is formed when the molecules are oriented to a common direction represented by a unit vector, n, or director. Hence, optical parallelism that depends on temperature is expressed quantitatively by order parameters S, which measures the degree of alignment of molecules' symmetry axes:

$$S = \frac{1}{2}\left(3\cos^2\theta - 1\right) \tag{1}$$

where θ is the angle between the director n and the local molecular orientation, while ⟨⋯⟩ indicates the thermal average.

Subsequent discovery of a new type of nematic phase led to its classification into two subclasses: nematic uniaxial N_u (long-distance order in one preferred direction, when the director n does not distinguish between head and tails) [7] and biaxial nematic N_b, characterized by three orthogonal directors (a primary director n and two secondary directors l and m) [8]. Hence, the biaxial nematic phase has three different indices of refraction along the three spatial directions. In this case, the order parameter is better described by an asymmetric real tensor order parameter ~Q [9]:

$$\overline{Q} = \begin{bmatrix} -\frac{S+\eta}{2} & 0 & 0 \\ 0 & -\frac{S-\eta}{2} & 0 \\ 0 & 0 & S \end{bmatrix} \quad (2)$$

In Eq. (2), η measures the degree of biaxial order along the secondary director, m. When η is zero, it describes the uniaxial N_u phase, when S reduces to Eq. (1).

Confusions between molecular biaxiality and phase biaxiality are often created. In addition to the molecular aspect, the interactions between molecules, steric factors, dipoles, and the ordering anisotropy in the liquid crystalline structures are of decisive importance in observing the phase biaxiality. However, the higher the molecular biaxiality, the greater the chances of observing the phase biaxiality. Molecular modeling has played an important role in the study of the biaxial nematic; many papers, starting from mathematical considerations and based on potential models, deal with the simulation of molecular interactions and predict the observation of biaxiality in chosen systems. These models led to the identification of essential physical features, such as anisotropy, biaxiality, and electrostatic moments, and aimed at simulating bent-core systems [10] or disc/rod mixture systems or bent-core/rod mixture [11].

For a uniaxial nematic, the director can be aligned by a magnetic or electric field, so almost all liquid crystals display devices that used a uniaxial nematic phase. However, the biaxial nematic phase is technologically more important to exploit because of high sensitivity to magnetic or electric fields, flexoelectricity, or unusual rheological properties [8]. Hence, biaxial nematic liquid crystals can be successfully used for electro-optic switching or to control birefringence, considering negative dielectric anisotropy or lower viscosities associated with E-field reorientation of the transverse director about an aligned director n. Potential applications of bent-core nematic liquid crystals target optical data storage, holographic media, photonics, or photoalignment of LC matrices [12, 13].

The alignment and switching of mesophases by an external field depend on polarization of liquid crystals and orientation of the dipole (parallel to the molecular axis or parallel to the column axis respectively). Liquid crystals that are responsive to electric fields show ferroelectric and antiferroelectric states. The main characteristic of ferroelectric liquid crystals (FLC) is the spontaneous polarization. Compared to uniaxial nematic phases that do not exhibit polar properties, considerable interest for technological applications represents the polar switching of biaxial nematic phases and fast response speed at low applied field strength, respectively (microseconds vs. milliseconds for achiral nematics) [14].

If the polar order in polar smectic phases arises from chirality or asymmetric close-packing constraints, in the case of biaxial nematics, ferroelectricity is the result of dipole–dipole interactions (the chiral polar phases are formed even in the case when the bent-core molecules are non-chiral).

The polar behavior of switchable phase was predicted by molecular dynamics simulations, which identify the structure–property relationships and evidenced the detailed perspective of the molecular organization in liquid crystalline phases. Hence, the nature of biaxial nematic consists of nanosized clusters of molecules or cybotactic groups [15].

2. Molecular shape of nematic mesogens

Materials that exhibit liquid crystalline phases are called mesogens. The relationship between the microscopic shape of mesogens that form the nematic phase and the macroscopic symmetry of the phase affect their physical properties. Hence, when designing a molecular structure, the aim is to influence and particularly enhance the molecular polarization.

The uniaxial nematic phase is found in simplest low molar mass liquid crystals or mesophases represented by compounds made up of long cylindrical-shaped molecules (calamitic mesogens) (**Figure 1a**) or discs (discotic mesogens) (**Figure 1b**); the biaxial nematic phase is found to be prevalent in bent-core compounds (banana mesogens) (**Figure 1c**).

The general structure of nematic liquid crystal compounds is composed of aromatic cycles which are planar, rigid, and polarizable, connected by conjugated double bonds along the long axis of the molecule. The rigidity of calamitic (rod-shaped) molecules provides the orientational order of the molecular axes, while their functionalization with one or two flexible hydrocarbon tails provides mobility by preventing these systems from crystallizing. Overall, an essential condition that must be met is the length of the molecule that should be greater than its diameter. The mesomorphism is generally enhanced by the presence of strong polar groups near the center of the molecule and by the weak polar groups at the extremities of the molecule.

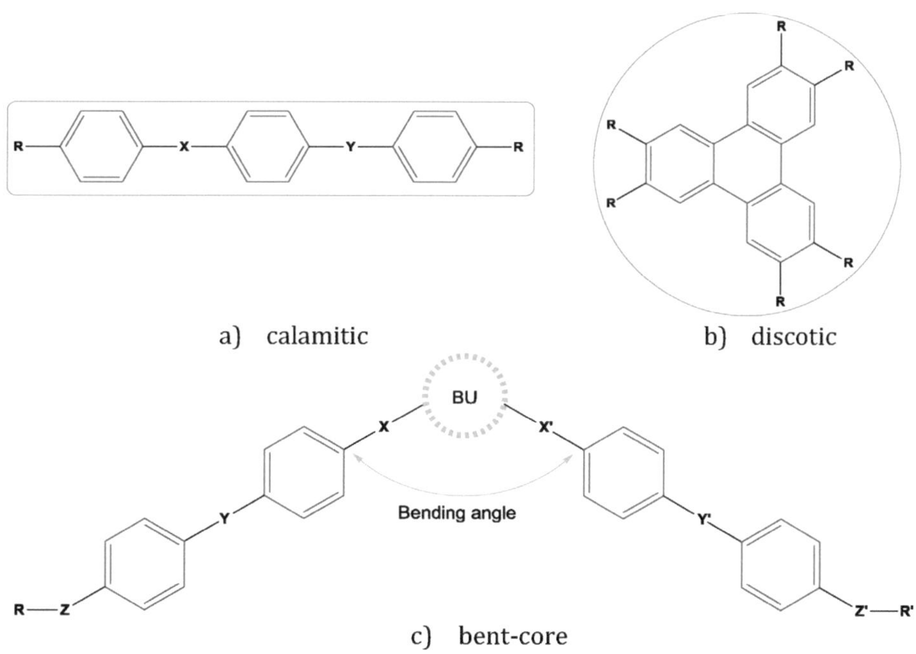

Figure 1.
Typical structures in thermotropic nematic liquid crystals: (a) calamitic, (b) discotic, and (c) bent-core.

If molecular units from nematic mesogens are chiral molecules or mixtures of achiral molecules with optically active compounds, the molecular orientation of a nematic phase is distinguished by a helical modulation, a characteristic to the cholesteric phase, closely related to the nematic phase. They are of great technical importance for appliances with displays.

The length of bent-core molecules and the bending angle of the central unit are most important for the formation of polar ordered mesophases. Hence, a minimum number of rings between four and five are required (for three rings the system is stabilized by intramolecular H-bonding). Symmetric bent-core units (BU) comprising 2,7-disubstituted naphthalene derivatives show mesophases with highest thermal stability, while less symmetric biphenyl derivatives exhibit wide mesomorphic domains. Generally, increasing the size of the rigid bent unit increases the transition temperatures, so in order to obtain polar phases, longer terminal chains are required. The nature, position, and direction of the linking groups between the rings influence as well the ordering in mesophases; the most commonly used are ester and a combination of ester and imine groups. The obtaining of polar mesophases depends as well on the bending angle, which can vary between 105 and 140°.

In contrast to classical nematics formed by calamitic mesogens, nematic mesophases formed by bent-core mesogens exhibit unique properties, due to high packing density in a uniform direction and a polar order in the layers [15, 16]. As a result, they exhibit mesophases with a permanent polarization in the absence of an electric field (spontaneous symmetry breaking of achiral molecules and super helical structures).

For most rigid bent-core mesogens, nematic phases are rarely observed because of high tendency to associate into mesophases with positional long-range order. By contrast, flexible bent-core molecules form nanostructured nematic phases, including the twist-bend nematic phase discovered very recently [17]. Accordingly, the bent core nematic phase contains uniaxial, biaxial, and polar nematics and phases with tetrahedratic symmetry [18].

3. Characterization of nematic mesophases

The most common techniques for characterization of nematic mesophases are polarized optical microscopy (POM), differential scanning calorimetry (DSC), X-ray scattering at wide and small angles (WAXS, SAXS), solid-state NMR, IR, and Raman spectroscopy [19].

Nematic phase is best highlighted in polarized optical microscopy, through which its fluidity and Schlieren texture are revealed (**Figure 2**). The method consists in observing the texture defects for which it was concluded that the presence of only the so-called two-brush disclinations is specific to the biaxial nematic mesophase and the presence of four-brush disclinations is characteristic of the uniaxial nematic mesophase. However, texture alone is not sufficient to determine whether the phase is uniaxial or biaxial. The microscopic method is considered less eloquent due to the influence of the sample preparation on the obtained image, surface effects, and the interaction between the liquid crystal sample molecules with the walls of the microscopic plate. It is used as an early method in identifying biaxial mesophase, but it must be accompanied by other methods in which no disruptive external factors are involved (e.g., solid-state NMR).

Conoscopy with and without a circular polarizer represents a valuable method to confirming biaxiality. This consists in looking at the interference image of a sample in the focal plane. The sample must be well aligned with the director n (homeotropic alignment) and oriented parallel to the incident light. The conoscopic

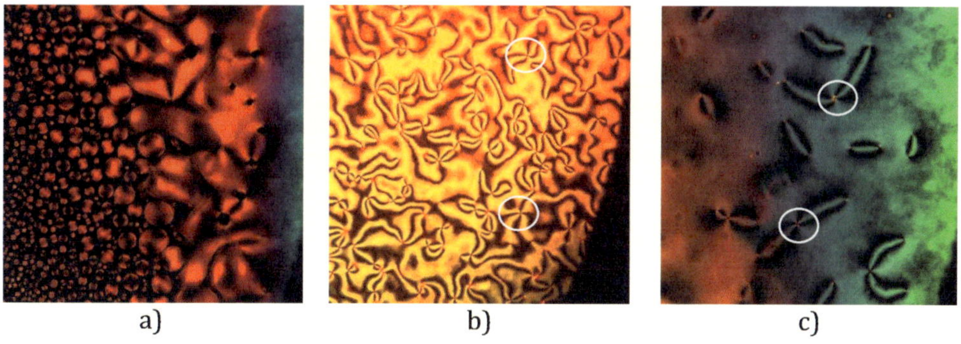

Figure 2.
(a) Nematic droplets, (b) Schlieren texture of a presumably uniaxial nematic (four-brush disclinations), and (c) Schlieren texture of a presumably biaxial nematic phase (two-brush disclinations).

image is characterized by the presence of so-called conoscopic isogyres that provide information about the axiality of the sample [8]. If the mesophase is uniaxial, the Maltese isogyre is observed, while if it is biaxial, a separation of the isogyres is observed, obtaining a characteristic conoscopic image.

Texture defects, singularities, or disclinations appear in the topology (arrangement of molecules) of a mesophase and are quantified by the factor S. The factor S is defined macroscopically, for the nematic mesophase, by the number of curves without birefringence (dark brushes) that meet in a point (S = no. of curves/4) and can have the values ±1 (for uniaxial nematic) or ±1/2 (biaxial nematic) (**Figure 2**).

Differential scanning calorimetry signals, associated transition enthalpies and entropies, represent the first indications that a phase might be of special interest. Hence, the first-order nature of the transitions between crystalline, liquid crystalline, and disordered phase is evidenced. However, the second-order transitions of conventional uniaxial nematic phases N_u or polar biaxial N_b phases are difficult to detect only using the DSC technique, which gives only a clue of the existence of a mesophase; therefore a combination with other methods is required.

X-ray diffraction represents the most important method of identifying mesophases. The analysis of the position and intensity of X-ray reflections allows the complete description of the mesophase structure. For nematic phases, postulates is the proof of the presence of cybotactic clusters in the diffractogram of phases as related to biaxiality.

Additional analytical methods for characterizing mesophases include miscibility studies with known liquid crystals, electro-optical measurements, or light scattering to test for cluster formation in uniaxial nematics or ESR spectroscopy, dielectric spectroscopy, atomic force microscopy (AFM), and rheology.

Author details

Irina Carlescu
Faculty of Chemical Engineering and Environmental Protection, Iași, Romania

*Address all correspondence to: icarlescu@ch.tuiasi.ro

References

[1] Demus D, Goodby J, Gray GW, Spiess HW, Vill V. Nematic Liquid Crystals: Applications, Section 3.1. Weinheim, Germany: Wiley-VCH Verlag Gmb; 1998. DOI: 10.1002/9783527620555.ch3h

[2] Zhao JG, Gulan U, Horie T, Ohmura N, Han J, Yang C, et al. Advances in biological liquid crystals. Small. 2019;**15**(18):1-12. DOI: 10.1002/smll.201900019

[3] Lowe AM, Abbott NL. Liquid crystalline materials for biological applications. Chemistry of Materials. 2012;**24**(5):746-758. DOI: 10.1021/cm202632m

[4] Kim YK, Noh J, Nayani K, Abbott NL. Soft matter from liquid crystals. Soft Matter. 2019;**15**(35):6913-6929. DOI: 10.1039/c9sm01424a

[5] Miller DS, Wang X, Abbott NL. Design of functional materials based on liquid crystalline droplets. Chemistry of Materials. 2014;**26**(1):496-506. DOI: 10.1021/cm4025028

[6] Demus D. Liquid Crystals. Darmstadt: Verlag; 1994

[7] Blinov LM. Structure and Properties of Liquid Crystals. Dordrecht/Heidelberg/London/New York: Springer; 2011

[8] Luckhurst GR, Sluckin TJ. Biaxial Nematic Liquid Crystals: Theory, Simulation, and Experiment. West Sussex, United Kingdom: John Wiley & Sons; 2015

[9] de Gennes PG, Prost J. The Physics of Liquid Crystals. New York: Oxford University Press; 1993

[10] Francescangeli O, Stanic V, Torgova SI, Strigazzi A, Scaramuzza N, Ferrero C, et al. Ferroelectric response and induced biaxiality in the nematic phase of a bent-core mesogen. Advanced Functional Materials. 2009;**19**(16):2592-2600. DOI: 10.1002/adfm.200801865

[11] Kundu B, Pratibha R, Madhusudana NV. Orientational order in liquid crystals exhibited by some binary mixtures of rod-like and bent-core molecules. European Physical Journal E: Soft Matter and Biological Physics. 2010;**31**(2):145-152. DOI: 10.1140/epje/i2010-10556-8

[12] Etxebarria J, Ros MB. Bent-core liquid crystals in the route to functional materials. Journal of Materials Chemistry. 2008;**1**(25):2919-2926. DOI: 10.1039/b803507e

[13] Bisoyi HK, Li Q. Light-driven liquid crystalline materials: From photo-induced phase transitions and property modulations to applications. Chemical Reviews. 2016;**116**(24):15089-15166. DOI: 10.1021/acs.chemrev.6b00415

[14] Takezoe H. Polar liquid crystals—ferro, antiferro, banana, and columnar. Molecular Crystals and Liquid Crystals. 2017;**646**(1):46-65. DOI: 10.1080/15421406.2017.1284377

[15] Vita F, Adamo FC, Francescangeli O. Polar order in bent-core nematics: An overview. Journal of Molecular Liquids. 2018;**267**:564-573. DOI: 10.1016/j.molliq.2018.02.084

[16] Jakli A. Liquid crystals of the twenty-first century-nematic phase of bent-core molecules. Liquid Crystals Reviews. 2013;**1**(1):65-82. DOI: 10.1080/21680396.2013.803701

[17] Mandle RJ. Designing liquid-crystalline oligomers to exhibit twist-bend modulated nematic phases. Chemical Record. 2018;**18**(9):1341-1349. DOI: 10.1002/tcr.201800010

[18] Lubensky TC, Radzihovsky L. Theory of bent-core liquid-crystal phases and phase transitions. Physical Review E. 2002;**66**:031704. DOI: 10.1103/PhysRevE.66.031704

[19] Goodby JW, Collings PJ, Kato T, Tschierske C, Gleeson H, Raynes P, et al. Handbook of Liquid Crystals, 8 Volume Set, 2nd Edition—Vol 1: Fundamentals of Liquid Crystals, Part III—Characterization of Liquid Crystals. Weinheim, Germany: Wiley-VCH; 2014

Chapter 2

An Overview of Polymer-Dispersed Liquid Crystal Composite Films and Their Applications

Anuja Katariya Jain and Rajendra R. Deshmukh

Abstract

Inherent and incredible properties of liquid crystals (LC) such as optical and dielectric anisotropy make them special candidates for flat-panel display devices; bi-stable reflective displays; high-definition spatial light modulators; switchable windows; haze-free normal- and reverse-mode light shutter devices; projectors; optical, thermal and strain sensors; tuneable lenses; etc. Non-linear response of LC material to the applied electric field is very useful in the above-mentioned applications. When a low molecular weight LC material is doped in a high molecular weight polymer matrix to obtain polymer-dispersed liquid crystal (PDLC) films, it offers flexibility and mechanical strength (structural stabilization) to the composite films—PDLC devices. Depending upon the concentration of monomer/polymer, these composite films are classified as polymer-stabilized liquid crystal (PSLC), PDLC and holographic PDLC (HPDLC) films. Depending upon the process conditions, we get phase-separated randomly dispersed micron-sized LC droplets in a continuous polymer matrix. These nematic LC droplets exhibit light scattering transmission properties depending on their orientation, which can be controlled by external electric field. This chapter gives deep insight about operating principle, phase separation techniques involved, alignment of LC and controlling LC droplet morphology of PDLC films to obtain desired properties. In order to improve the optical efficiency and to obtain the desired result from PDLC films, various guest entities such as dye and nanomaterials are doped in the host LC material. This chapter also accounts for various possible LC dopants desired for improving the electro-optic (EO) and dielectric properties of PDLC devices. Various applications of PDLC composite films are also described in this chapter.

Keywords: nematic liquid crystal, polymer-stabilized liquid crystal, polymer-dispersed liquid crystal, holographic polymer-dispersed liquid crystal, phase separation techniques, electro-optic and dielectric properties of PDLC

1. Definition and history of liquid crystals

Liquid crystal (LC) is a thermodynamic phase of a condensed matter, intermediate of (in between) conventional isotropic liquid and three dimensionally ordered solid crystal with only orientational order but no positional order [1]. It holds properties of liquid such as fluidity, coalescence and formation of droplets as well as crystalline properties such as order and anisotropy in optical, electrical and magnetic properties (as summarized in **Table 1**) [2]. The difference between molecular arrangement of solid, liquid crystal and liquid is shown in **Figure 1**.

This phase of matter was discovered by the Austrian botanist Friedrich Reinitzer in the year 1888 while he was studying the compounds cholesteryl benzoate and cholesteryl acetate. He observed colored phenomena occurring in melts of cholesteryl acetate and cholesteryl benzoate. In addition, he reported that the compound cholesteryl benzoate has two distinct melting points. Its crystal transforms into hazy liquid at a temperature 145.5°C, and with the further increase in temperature, it suddenly turns into isotropic liquid at a temperature 178.5°C [3]. The word "liquid crystal" for this unusual phase of material was coined by German Physicist, Otto Lehmann, a specialist in polarizing optical microscopy [4]. Friedrich Reinitzer's and Otto Lehmann's studies revealed that liquid crystals (LCs) can rotate the direction of polarization of light and reflect circularly polarized light [5]. The history of the development of LCs can be divided into three phases: the first phase is from the discovery of LC (1888) to the acceptance of its existence (1925). Friedel's article about classification of LC, along with the publications on synthesis and studies of new LC materials by organic chemists in Germany, notably Vorlander [6, 7], provided a firm basis for the development of the subject. In the period from 1925 to about 1960, research in the field of LC was at a low level; however contribution of some devoted researcher has been summarized here: Vorlander synthesized a number of compounds forming LC phases, some of them showing up to three different mesophases. A lamellar and tilted lamellar structure was found by Herrmann for some thallium soaps [8]. In this period, the effect of external electric or magnetic field on the LC has been recognized, modifications of surface to orient LC has been done, and significance of anisotropic physical properties of aligned LC has been understood, which was the foundation of display technology of the future world. Synthesis of new LC materials and their structure–property

Solid	Liquid crystal	Liquid
Anisotropic	Anisotropic	Isotropic
Rigidity	Fluidity	Fluidity
Ordered	Ordered	Disordered
3-D lattice	0/1/2 lattice	No lattice

Table 1.
Properties of solid, liquid crystal and liquid.

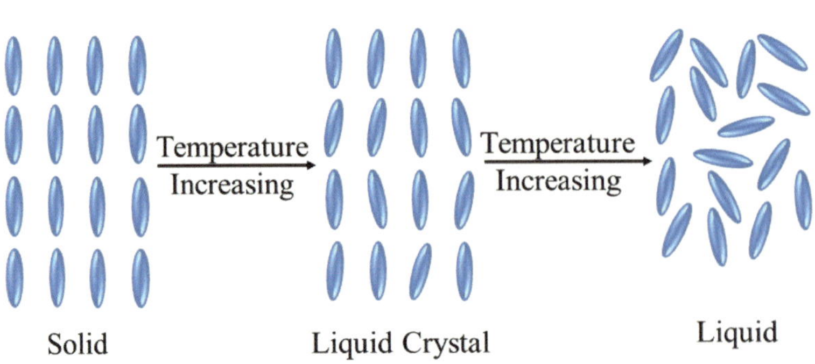

Figure 1.
Molecular arrangement of solid, liquid crystal and liquid.

relationships has been explained using principles commencing from swarm theory to continuum theory. Identification of mesophases, determination of transition temperatures and awareness about defect textures through polarizing optical microscope (POM) appreciably helped in developing a new era of LC applications [9]. The beginning of the third period (1960 to the present time) gives us a famous statistical Maier and Saupe mean field theory which states about the isotropic-nematic phase transition [10, 11]. Synthesis of new LC materials is ongoing. Classification, molecular structure and properties of LCs, defects characterizing microscopic textures of LC phases and existence of blue phases are some topics, which are now well understood [12–14]. The decade of 1970 is known for announcing application of LCs as display devices. Expansion of theories and their employment in practical applications give rise to new LC science from display world to beyond display technology as well. Having high resolution and high brightness and being lightweight, flexible and an energy saver make LC devices attractive and competitive in the high-tech world. LC science has been now well acknowledged and documented, but it is still thriving. Liquid crystal displays (LCDs) are like a milestone for the future world, but a continuous evolution via dedicated research is still anticipated.

2. Types and phases of liquid crystals

This special class of materials are moderate-sized organic molecules, composed of flat segments like benzene rings, double bonds, strong dipole and easily polarizing groups [15]. Depending upon the molecular structure, LC compounds may have one or many phases (polymorphism), characterized by order and symmetry. LCs can be broadly classified into two generic classes: thermotropic and lyotropic [1]. Commonly found LCs from either group possess a remarkable polymorphism and give rise to various mesophases such as nematic, smectic, cholesteric, columnar and blue phases in thermotropic LC and discontinuous, hexagonal, lamellar, bicontinuous, reverse hexagonal and inverse cubic phases in lyotropic LC, depending upon the type, amount and proportion of ordering in it [16]. Sometimes, thermotropic LCs are also cataloged as rod-like (calamatic) and disk-like (discotic). Thermotropic LC are single compounds, whereas lyotropic LC are always mixtures. The LiqCryst database accounts for more than 39,000 nematic phase compounds, about 18,000 chiral compounds and more than 6000 ferroelectric SmecticC* phase compounds [17]. Some day-to-day life examples of LCs are soap solution, tobacco mosaic virus, protein and cell membrane.

2.1 Thermotropic liquid crystals

Thermotropic LCs are comprised of rod-like organic molecules and exhibit phase transition into the LC phase as a function of temperature. At very low temperature, most LC materials are in anisotropic phase, but with the increase in temperature, these LC materials acquire isotropic phase along with so many intermediate phases such as smectic, nematic, cholesteric etc., which are described below [1, 15].

2.1.1 Nematic phase

As the temperature of isotropic phase (no positional or orientation order) is lowered, the LC material undergoes a transition to the nematic phase. It is a transparent or translucent low-viscosity liquid and a stable LC phase in a particular temperature range. It is the most common LC phase of calamatic or rod-shaped organic molecules, as shown in **Figure 2(a)** [4].

The structure of a typical nematic LC is shown in **Figure 2(b)**, and each entity has exclusive function. The terminal group (e.g. $(C_6H_{13})_n$) determines dielectric constant and anisotropy, and benzene rings provide short-range molecular forces which affect electrical and elastic properties; the linkage group stabilizes LC against moisture, UV radiation and chemicals, and the side chain (e.g. cyano group) influences the elastic constants and transition temperature of LC. **Figure 2(c)** represents example of nematic LC. Nematic LCs lack positional order, but have self-aligning long-range directional order with their long axes almost parallel, characterized by a nematic director $\hat{n}$, which is the average direction of the ensemble of molecules [15]. The director $\hat{n}$ is a function of space with unit magnitude and $\hat{n} = -\hat{n}$. Thus, the LC molecules in nematic phase are free to flow with three translational degrees of freedom, and their centre of mass positions is randomly distributed as in a liquid, but still maintains their long-range directional order. In nematic phase of LC, one axis is generally longer and preferred than the other two, i.e., they are uniaxial and can be approximated as cylinders and rods. The easy alignment of these uniaxial nematic LC using electric or magnetic field makes them optically uniaxial and prominent in display devices. However, some LCs are biaxial nematic, i.e., in addition to orienting along their long axis, they also orient along a secondary axis [1, 18]. As the nematic LC is relatively a low-viscosity fluid, it easily gets deformed by small external forces. In a deformed LC, the director direction $\hat{n}$ changes from point to point. These LC deformations can be explained using three basic deformations: splay, twist and bend,

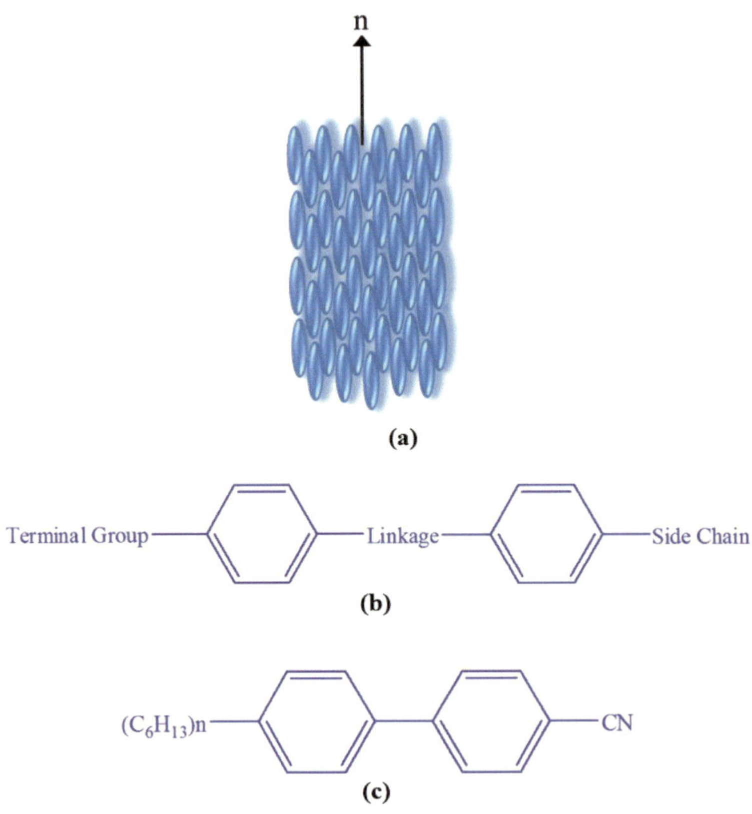

Figure 2.
(a) Molecular arrangement, (b) general chemical structure and (c) example of nematic liquid crystal.

and their associated elastic constants are K_{11}, K_{22} and K_{33}, respectively. The free energy of distortions of nematic LC is given by Eq. (1) [19]:

$$F = \frac{1}{2}\left[K_{11}\left(\nabla.\vec{n}\right)^2 + K_{22}\left(\vec{n}.\nabla \times \vec{n}\right)^2 + K_{33}\left(\vec{n} \times \nabla \times \vec{n}\right)^2\right] \tag{1}$$

2.1.2 Cholesteric phase

The cholesteric LC phase is typically composed of nematic mesogenic molecules containing chiral centres. Chiral molecules have no internal planes of symmetry and produce a twist in the nematic structure by inducing intermolecular forces which favour alignment between molecules at a slight angle to one another. They are composed of quasi-nematic layers. Their individual directors are turned by a fixed angle on proceeding from one layer to the next as shown in **Figure 3(a)**. The rotation is constrained in a plane perpendicular to the pitch direction. The pitch (p) is the distance over which the director of LC molecules undergoes a full twist of 2π angle. As the phase directors at 0° and 180° are equivalent, the arrangement of molecules in the chiral nematic phase repeats at every half pitch (p/2). Due to this strong twisting effect, in a definite spectral range, cholesteric phase shows a selective reflection of the circularly polarized light of wavelength equal to pitch length. The pitch length p can be altered by varying temperature or adding other materials in LC host. With the increase in temperature, the angle at which the director changes increases, which in turn decreases the pitch length and vice versa [4, 20, 21]. The free energy of distortions in cholesteric LC is given by Eq. (2)

$$F = \frac{1}{2}\left[K_{11}\left(\nabla.\vec{n}\right)^2 + K_{22}\left(\vec{n}.\nabla \times \vec{n} + q_0\right)^2 + K_{33}\left(\vec{n} \times \nabla \times \vec{n}\right)^2\right] \tag{2}$$

where $q_0 = \frac{2\pi}{p}$ corresponds to the intrinsic twist of the system. For nematics p is infinite; therefore q_0 vanishes from free energy equation of nematic LC [22]. **Figure 3(b)** represents example of cholesteric LC.

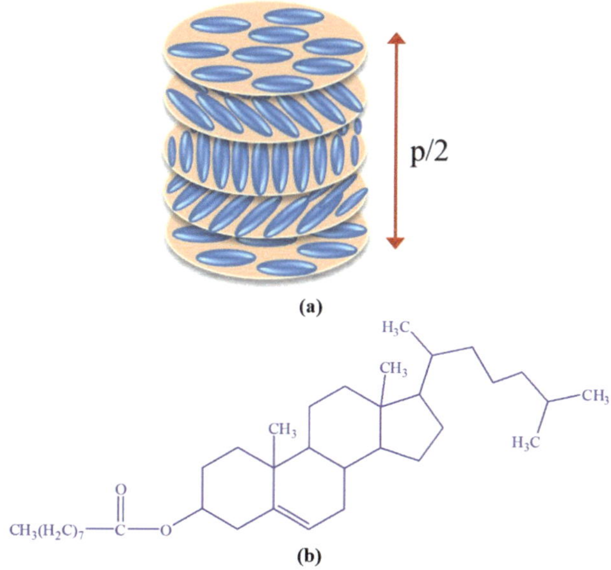

Figure 3.
(a) Molecular arrangement and (b) example of cholesteric LC phase.

2.1.3 Blue phase

The blue phases are a set of thermodynamically distinct phases that occur at the boundary of the helical phase and isotropic phase of highly chiral LCs within a small temperature range. It was first observed by Reinitzer in 1888 as an unstable phase, and after a century (in 1975), they were shown to be stable and distinct thermodynamic phase by Armitage and Price [23, 24]. In the absence of electric fields, in the order of increasing temperature, there can be three blue phases: BPI*, BPII* and BPIII*. BPI* has body-centred cubic symmetry, BPII* possess simple cubic symmetry and BPIII* is with a local cubic lattice only. BPI* and BPII* reflect blue light, as their name suggest, whereas BPIII* phase is observed at highest temperature and appears fogy because of which it is called as fog phase or blue fog [25, 26]. The building structure element of BPI* and BPII* phase is double-twist cylinders (**Figure 4(a)**). The double-twist cylinder is a local structure of minimum free energy with local director rotating around any given radius of the cylinders. Free energy of blue phases is lower than the free energy of chiral nematic phase because here molecules twist in two dimensions simultaneously (**Figure 4(b)**). As the local twist is increased, the cylinder becomes strained and distorted. Therefore, blue phase cannot have a single, large double-twist structure; instead it consists of many of these double-twist structures arranged in a lattice with cubic symmetry. But for elastic reasons, it is only possible by introducing a lattice of topological defects [27, 28] as shown in **Figure 4(c)**.

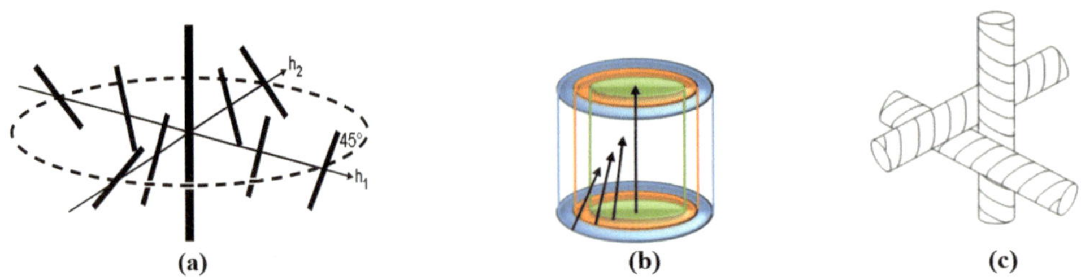

Figure 4.
(a) Double twisted structure with two helical axes, h1 and h2 in a blue phase LC, the directors perform a rotation of 90° across the diameter. (b) Perspective view of the double twist cylinder, the angle of directors at the outer edge of the cylinder is 45°, relative to the central axis. (c) Local arrangement of three double twist cylinders forming a defect region, which eventually leads to the three-dimensional, cubic lattice of defects observed in blue phases.

2.1.4 Smectic phase

Upon cooling, nematic phase LC transforms into smectic phase. The distinguished feature of smectic phase LC is their stratification. In addition to the orientational order, their molecules form well-defined layers which can slide over one another. Thus, they are positionally ordered along one direction with two translational degrees of freedom. The increased order indicates that the smectic phase is more solid-like than the nematic. Several different smectic classes (phases) have been discovered so far, and some of them are discussed here. In the smectic A (SmA) phase, on average, the molecules are parallel to one another possessing orientational order and are arranged in layers, with the long axes perpendicular to the layer plane. The orientational order is characterized by the director $\hat{n}$ analogous to the nematic LC but restricted within a specific layer/plane. Within the layers the centre of mass of the molecules is ordered at random and has no correlation between intra-plane centres of masses. Thus, the SmA phase (**Figure 5(a)**) possesses the one-dimensional quasi long-range positional order, and within the layers,

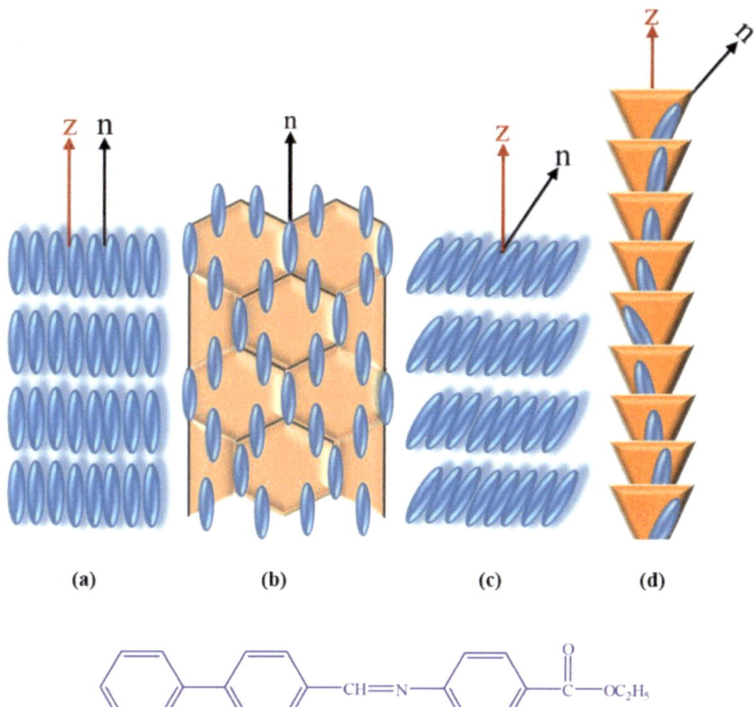

Figure 5.
Molecular arrangement in (a) smectic A phase, (b) smectic B phase, (c) smectic C phase, (d) smectic C phase and (e) example of smectic phase.*

molecules show relatively high mobility. The thickness of layer is equal to molecular length. The SmA LCs are optically positive and uniaxial with the optic axis parallel to the molecular long axes. The layers of the SmA LC can be bended in a way causing splay deformation. Bend and twist deformations are prohibited in this LC phase. The free energy density equation is given by Eq. (3)

$$F = \frac{1}{2}\left[K_{11}\left(\nabla.\vec{n}\right)^2\right] \tag{3}$$

Upon further cooling, the smectic B (SmB) (**Figure 5(b)**) and smectic C (SmC) (**Figure 5(c)**) phases are formed. The SmB mesophase orients with the director perpendicular to the smectic plane, but the molecules are arranged into a network of hexagons within the layer. In the SmC mesophase, molecules are arranged as in the SmA mesophase, but the director is at a constant tilt angle measured normally to the smectic plane. For some material the tilt angle is constant, but for others it is temperature dependent. The centre of mass of the molecules is randomly oriented/ordered, and the molecules are free to rotate around their long axes. SmC phases are optically biaxial. If the molecules of SmC LC are in chiral state, then they are designated as smectic C* (SmC*) (**Figure 5(d)**) state, and the direction of the director projection is rotated from layer to layer forming a helix. Therefore, these phases appear optically positive uniaxial and show optical activity and selective reflection similar to the cholesteric phase. The SmC* shows ferroelectric properties if their molecules have permanent dipole moment perpendicular to their long axes.

In some smectic phases (e.g. Smectic G phase), the molecules are affected by the various layers above and below them. Therefore, a small amount of three-dimensional order is observed in them [1, 2, 15, 21]. **Figure 5(a–d)** shows the molecular arrangement in all types of smectic phases, and **Figure 5(e)** is an example of smectic phase.

2.1.5 Discotic phase

Apart from the rod-like molecules, more advanced-shaped LCs are possible such as disk-like (**Figure 6(a)**) which can give rise to other types of ordering. They were first discovered in carbon precursor compounds with a transient existence by Brooks and Taylor in stable low molecular weight systems [29, 30].

Disk-shaped LC molecules can orient themselves in a layer-like manner termed as the discotic nematic phase. This phase is called as a discotic columnar, if their disks pack into stacks/columns. Again, these columns may organize themselves into rectangular or hexagonal arrays [31]. Discotic LCs are composed of an aromatic core surrounded by flexible chains as shown in **Figure 6(b)**. The aromatic cores allow charge transfers in the stacking direction through the π conjugate system, due to which these LCs become electrically semiconducting along the stacking direction.

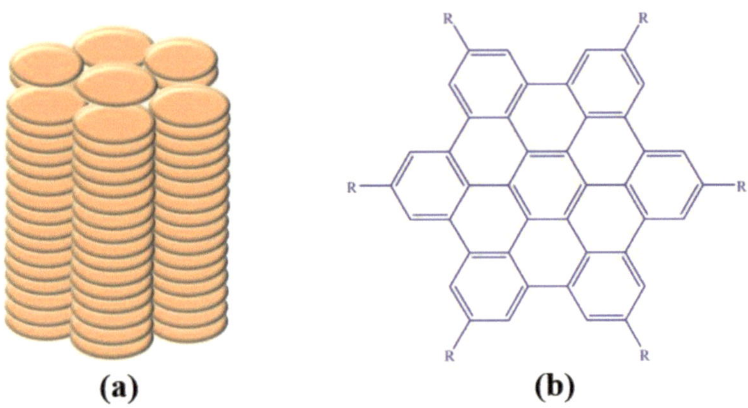

Figure 6.
(a) Molecular arrangement and (b) example of disk-shaped LC.

2.1.6 Banana-shaped LC

Sterically induced packing of bent core (banana-shaped) LC molecules (**Figure 7(a)**) is interesting from many viewpoints. These are the first ferroelectric and anti-ferroelectric LCs, which contain no chiral carbon atoms; however they can introduce chirality to the system [32]. One example of banana-shaped LC is shown in **Figure 7(b)**.

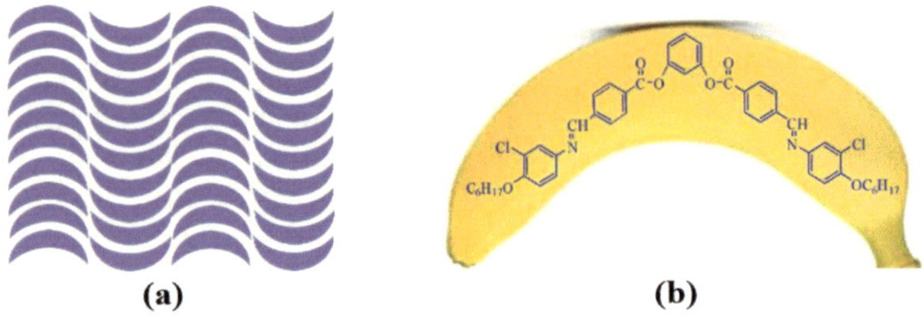

Figure 7.
(a) Molecular arrangement and (b) example of banana-shaped LC.

2.2 Lyotropic liquid crystals

Another class of LCs is named as lyotropic LCs, having two distinct parts/building blocks—hydrophobic and hydrophilic. Their properties depend on the concentration in the solvent and the shape of the molecule. Soaps and detergents are some common examples of lyotropic LCs. It consists of two or more components that exhibit phase transition into the LC phase as a function of both temperature and concentration of the molecules in a solvent (generally water). The solvent molecules fill the space around the compounds and provide fluidity to the system. In lyotropics, along with temperature, concentration is another degree of freedom that enables them to induce a variety of different phases. A compound which has two immiscible hydrophobic and hydrophilic parts within the same molecule is termed as an amphiphilic molecule. Depending on the volume balances between the hydrophobic part and hydrophilic part, many amphiphilic molecules show lyotropic liquid-crystalline phase sequences. These structures are formed because of the micro-phase segregation of two incompatible components on a nanometre scale. At very low amphiphile concentration, the molecules are randomly dispersed in a solvent without any order. At slightly higher concentration, amphiphilic molecules spontaneously assemble into micelles or vesicles. This is done to "hide" the hydrophobic tail of the amphiphile inside the micelle core, exposing a hydrophilic (water-soluble) surface to aqueous solution. However, these spherical objects do not order themselves in solution. At higher concentration, the assemblies are well ordered. An example of such phase is a hexagonal columnar phase (**Figure 8(a)**). In this phase, the amphiphiles form long cylinders (again with a hydrophilic surface) that arrange themselves into a roughly hexagonal lattice. This is called the middle soap phase. At further higher concentration, a lamellar phase (**Figure 8(c)**) (neat soap phase) may form. In this phase extended sheets of amphiphiles are separated by thin layers of water. For some systems in between the hexagonal and lamellar phases, a cubic phase (**Figure 8(b)**) (viscous isotropic) may exist. In this phase spheres are formed that create a dense cubic lattice. These spheres may also be connected to one another, forming a bicontinuous cubic phase. The objects created by amphiphiles are usually spherical (as in the case of micelles), but sometimes disk-like (bicelles), rod-like or biaxial (all three micelle axes are distinct) objects are also possible. These anisotropic self-assembled nanostructures can then order themselves in similar way as thermotropic LCs do, forming large-scale versions of all the thermotropic phases (such as a nematic phase of rod-shaped micelles). For some systems, at high concentrations, inverse phases are observed, i.e., one may generate an inverse hexagonal columnar phase (columns of water encapsulated by amphiphiles) or an inverse

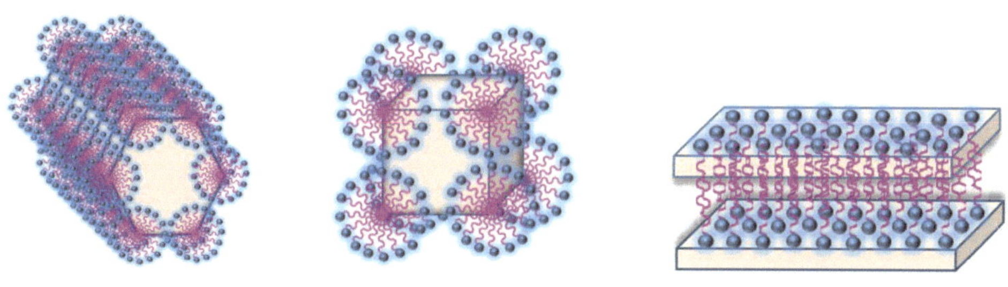

(a) (b) (c)

Figure 8.
Molecular arrangement of (a) hexagonal phase, (b) micellar cubic phase and (c) lamellar phase of lyotropic LC.

micellar phase (a bulk LC sample with spherical water cavities) [33, 34]. Different lyotropic phases are listed below:

1. Hexagonal phase (hexagonal columnar phase) (middle phase) (**Figure 8(a)**)
2. Discontinuous cubic phase (micellar cubic phase) (**Figure 8(b)**)
3. Lamellar phase (**Figure 8(c)**)
4. Bicontinuous cubic phase
5. Reverse hexagonal columnar phase
6. Inverse cubic phase (inverse micellar phase)

By varying concentration, even within the same phases, their self-assembled structures can be tuned. For example, in lamellar phases, distance between the layers increases with the solvent volume. Since lyotropic LCs indirectly depend on a subtle balance of intermolecular interactions, it is difficult to analyse their properties and structures as compared to those of thermotropic LCs. Similar type of phases and properties has been observed in immiscible diblock copolymers.

3. Properties of liquid crystals

3.1 Order parameter

To quantify amount of the orientational order in the LC phase, the term order parameter has been introduced; it is a second-rank symmetric traceless tensor defined as

$$S = \frac{1}{2}\left\langle 3\cos^2\theta - 1\right\rangle; \text{For nematic phase } 0.5 < S < 0.7 \tag{4}$$

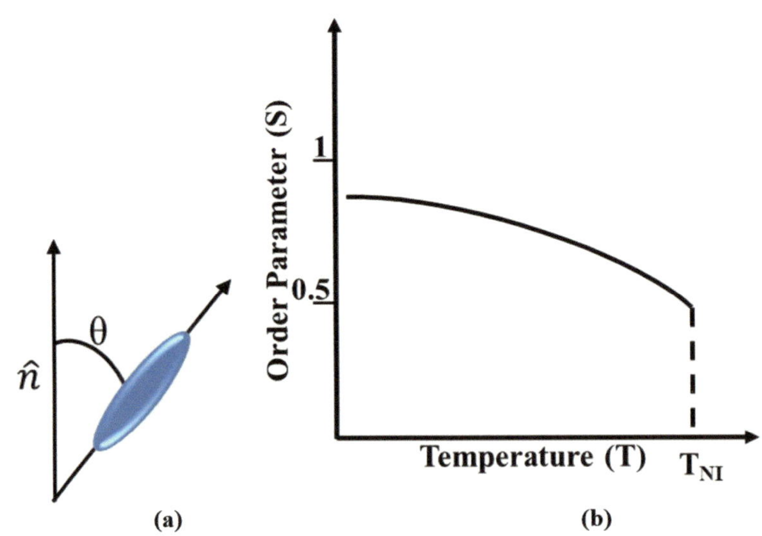

Figure 9.
(a) Liquid crystal director direction and (b) temperature dependence of order parameter.

where θ is the angle between the axis of an individual molecule and the local director $\hat{n}$ as shown in **Figure 9(a)**. It is the preferred direction in a volume element of a LC, and the average is taken over the complete ensemble. The bracket denotes both temporal and spatial average. For a completely isotropic sample, S = 0, whereas for a perfectly aligned sample, S = 1. For a typical LC sample, the value of S is 0.3 to 0.9, and for nematic LC, it is 0.5–0.7. **Figure 9(b)** shows the temperature dependence of order parameter (S), which follows an inverse relation [11, 35, 36].

3.2 Anisotropy in liquid crystals

LCs exhibit uniaxial symmetry around the director, which gives them shape anisotropy. The shape anisotropy of LC and their resulting interactions with the surrounding environment (applied fields) leads to an anisotropy in many other physical properties such as refractive index (RI), dielectric permittivity, magnetic susceptibility, viscosity and conductivity.

3.2.1 Optical anisotropy

LCs are optically anisotropic materials and show birefringence. LCs have two direction-dependent refractive indices, ordinary RI (n_o) and extraordinary RI (n_e) with birefringence:

$$\Delta n = n_e - n_o. \tag{5}$$

Also, the average RI is given by

$$n_{av} = \sqrt{\frac{1}{3}\left(n_e^2 + 2n_o^2\right)} \tag{6}$$

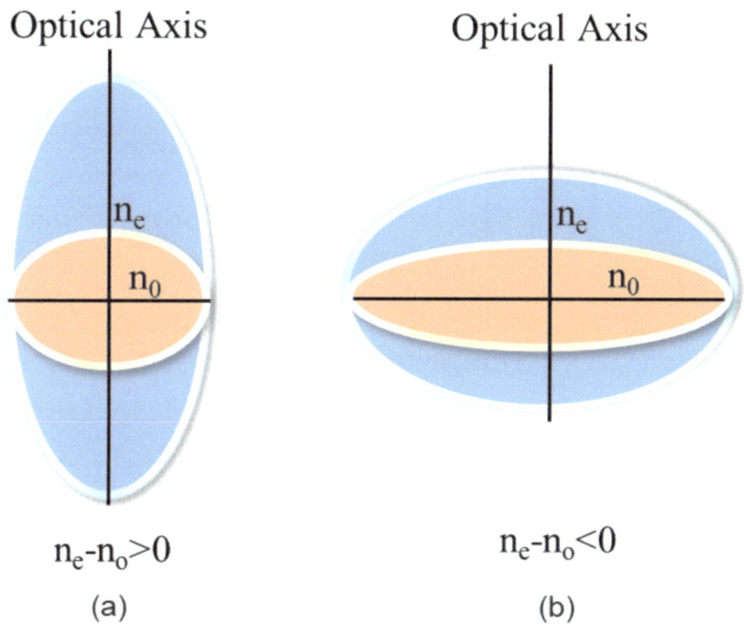

Figure 10.
Indicatrix of optically uniaxially material: (a) positive birefringent material and (b) negative birefringent material.

The value of Δn may be positive or negative, which can be represented by indicatrix as shown in **Figure 10**. For uniaxial crystal, it is ellipsoid where the rotational axis is identical to the optical axis [34, 37, 38].

For rod-like molecules (Nematic LC) $n_e > n_o$, where Δn is positive and between 0.02 and 0.4. For discotic and chiral nematic molecules $n_e < n_o$, and thus negative birefringence is associated with the discotic or columnar phase.

The values of optical anisotropy Δn can be increased [39]:

1. by replacing saturated aromatic rings with the unsaturated ones

2. with the elongation of the conjugation chain parallel to the long molecular axis

3. by increasing the values of the order parameter S or decreasing the value of temperature

4. by shortening the alkyl chain of the end molecular groups in homologous series in the form of even-odd alternation

In general, birefringence Δn of LCs decreases as the wavelength of the incident light or the temperature increases. Also, if the temperature of the LC material is raised up to its clearing point/nematic-isotropic temperature (T_{NI}), its internal order gets destroyed, and it behaves like an isotropic liquid with RI n_{iso} as shown in **Figure 11**.

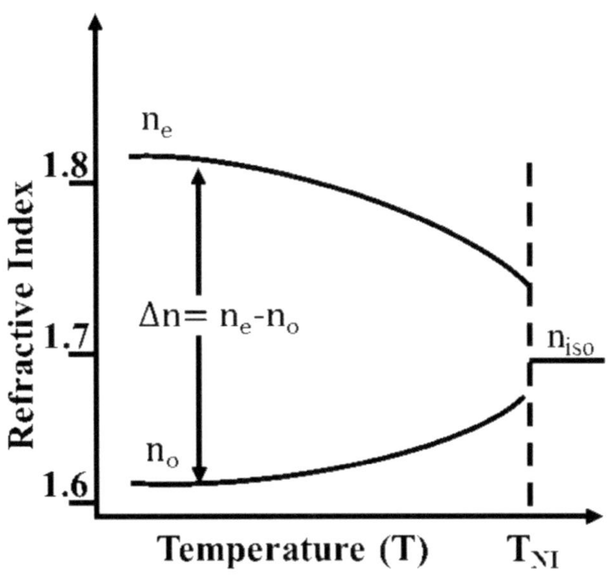

Figure 11.
Temperature dependence of refractive index (RI).

3.2.2 Dielectric anisotropy

Dielectric properties of LCs are related to the response of LC molecules upon application of an electric field. Permittivity is a property of a material that determines how dielectric medium affects and is affected by an electric field. It is determined by the capability of a material to polarize upon application of an electric field and in turn partially cancels the field induced inside the material [40, 41].

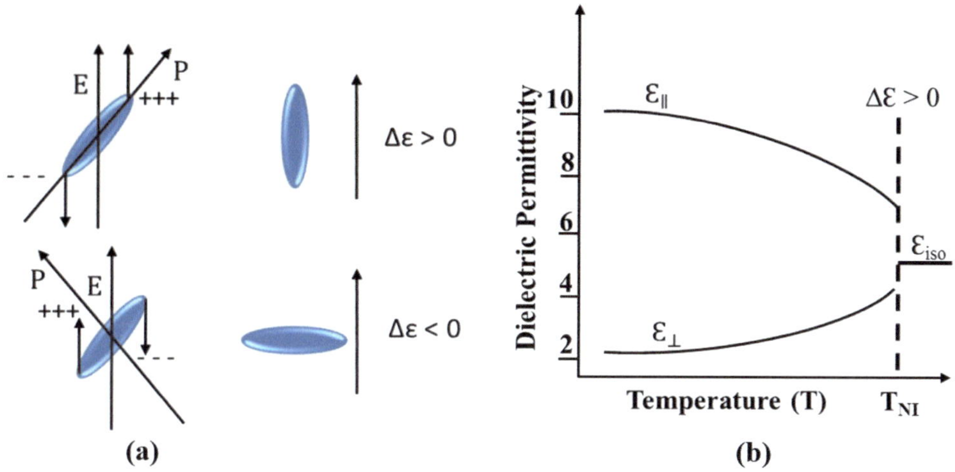

Figure 12.
(a) Alignment of positive and negative dielectric anisotropic LCs in external field, (b) temperature dependence of dielectric constant.

In the LC materials, consisting of non-polar molecules, there is only an induced polarization, which consists of two parts: the electronic polarization (which is also present at optical frequencies) and the ionic polarization. In the LCs with polar molecules, the orientational polarization exists along with the above-mentioned polarization. Considering the uniaxial LC phases in a macroscopic coordinate system, x, y and z, with the z-axis parallel to the director $\hat{n}$, it is possible to distinguish two principal permittivities, parallel to the director $\varepsilon_{/\!/} = \varepsilon_{zz}$ and perpendicular to the director $\varepsilon_{\perp} = 1/2(\varepsilon_{xx} + \varepsilon_{yy})$. $\varepsilon_{/\!/}$ is the characteristic of nematic LCs, as it corresponds to the polarization contribution related to the molecules. Then the dielectric anisotropy $\Delta\varepsilon = \varepsilon_{/\!/} - \varepsilon_{\perp}$ can take positive or negative values. If the value of $\Delta\varepsilon > 0$, then LC molecules align parallel to the field, whereas if the value of $\Delta\varepsilon < 0$, then the LC molecules tend to align perpendicular to the field (**Figure 12(a)**). The graph of temperature dependence of dielectric permittivity for a typical LC (**Figure 12(b)**) shows that magnitude of $\Delta\varepsilon$ usually depends on temperature. With the increase in temperature, liquid crystal material behaves as isotropic liquid with dielectric permittivity ε_{iso} [15].

The mean dielectric permittivity $\overline{\varepsilon}$ is temperature and frequency dependent, which can be described as

$$\overline{\varepsilon} = \frac{1}{3}\left(\varepsilon_{/\!/} - \varepsilon_{\perp}\right) \tag{7}$$

3.3 Elastic properties

The behaviour of LCs in an external electric field is highly dependent on their viscoelastic properties. While dealing with the elasticity of the nematic LC, we assume that the order parameter S remains invariable throughout the volume of LC at a constant temperature T and only director $\hat{n}$ changes with external field. The elastic constants of LCs associated with the restoring torques become apparent when the system is perturbed from its equilibrium configuration. These are of the order of 10^{-11} N (especially for nematic and fluid smectic phases), which suggests that a LC can be easily deformed by external forces, such as mechanical, electric or magnetic. The resistance of the LC to the external field gives rise to deformation. Final deformation pattern depends on the contribution of the associated elastic

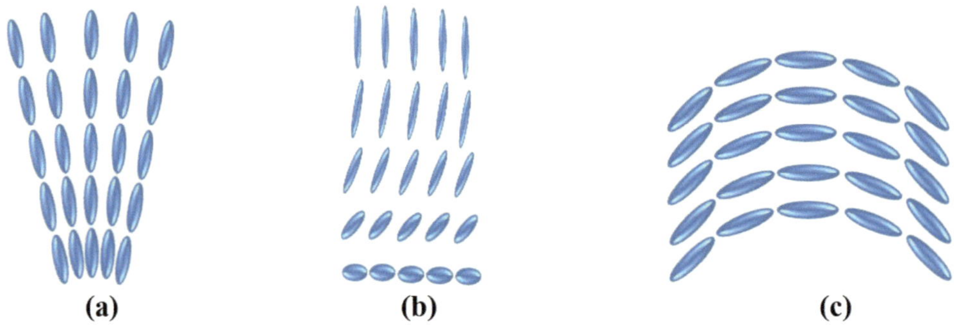

Figure 13.
(a) Splay, (b) twist and (c) bend deformation.

constants in elastic energy. For nematic LCs, it is assumed that change in elastic energy is only due to splay, twist and bend type deformation (**Figure 13**). The increase of free energy F due to these deformations is described by the continuum theory. This theory was first developed by Oseen and Zocher and later reformulated by Frank. It was based on the balance laws for linear and angular momentum [4, 42]. The contribution of each deformation to the overall energy F is given by

$$F = \frac{1}{2}\left[K_{11}(\nabla.n)^2 + K_{22}(n.\nabla \times n)^2 + K_{33}(n \times \nabla \times n)^2\right] \tag{8}$$

where K_{11}, K_{22} and K_{33} are proportionality constants of splay, twist and bend deformations, respectively, often known as Frank elastic constants [19]. They were forced to splay, twist and bend until equilibrium. When the system is in equilibrium, it is in minimum energy state [15]. Other types of deformation are forbidden due to the symmetry and absent polarity.

3.4 Viscosity

The dynamics of LC is described by (i) velocities of the centres of the molecules v and (ii) director field $\hat{n}$. Generally, these variables obey equation of continuity in incompressible liquids, Navier–Stokes equation in anisotropic viscous liquid and the equation of rotation of director in nematic LC. While dealing with the rotation of director, backflow effect should be considered, which states that the rotation of molecules (after removing external field) induces a macroscopic translational motion in LCs. However, the mathematics associated with the above-mentioned equations is insufficient to explain the viscosity behaviour of different LC substances and their mixtures. But in order to develop new liquid crystalline low-viscosity materials, the following phenomenological rules should be remembered [39]:

1. Alkyl end groups provide lower values of viscosity than alkoxy and acyloxy end groups.

2. The viscosity is lower for shorter molecules.

3. Introducing the rings with heteroatoms increases viscosity compared to phenyl analogues.

4. The most viscous bridging groups are the ester group –COO–, the simple bond (as in biphenyls) and the ethane group –CH=CH–.

5. Replacement of phenyl ring by a trans-cyclohexane ring results in reduced viscosity values.

The most useful compounds for reducing viscosity in LC materials are cyclohexane derivatives due to their high clearing temperature, good solubility and low viscosity.

3.5 LC in electric and magnetic field

The dependence of the free energy "*F*" of nematic LC on gradients of the director field is a unique property of LC. Therefore, the measurement of elastic constants in LC is a very crucial part in LC studies. The idea behind K_{ii} measurement is related to the registration of spatial distortions in structure induced by different factors such as electric field, magnetic field and thermal and surface fluctuations for which the following methods can be employed:

1. Optical method (Freedericksz transition)

2. Light scattering

3. Alignment inversion walls

4. Cholesteric-nematic transition

Out of which the optical method based on Freedericksz transition is the simplest and most significant from the application point of view [40].

3.5.1 Freedericksz transition

In the absence of any surface alignment or external field, LC directors of nematic molecules are free to point in any direction. However, it is possible to force the director to point in a specific direction by introducing an outside agent to the system. For example, when a thin layer of polymer (usually a polyimide (PI)) is coated on a glass substrate and rubbed in a single direction with a velvet cloth, it is observed that LC molecules in contact with that surface get aligned along the rubbing direction and achieve uniform director configuration. Upon application of magnetic or electric field for any distortion to occur (to overcome the elastic and viscoelastic forces of LC), the strength of the applied field has to be larger than certain threshold value [21]. Initially, when electric field is low, no change in alignment occurs. However, as we increase electric field above threshold, the LC director changes its orientation from one molecule to the next, and deformation occurs. This threshold is called the Freedericksz threshold, and the transition from a uniform director configuration to deformed director configuration is named as Freedericksz transition. To find out various elastic constants, we need to understand geometry of confined LC molecules and applied external field. The external field may be either electric or magnetic; it is more convenient and accurate to record electric field because measurement of magnetic field near to the sample is a tricky procedure due to the field inhomogeneity, and temperature dependence of Hall probe, etc.

Consider two, coated and rubbed (along *X* direction), conducting glass plates, separated by a distance "*d*". Due to this, LC director tends to align along the direction parallel to the flat surface (*X* direction). Now we consider the following cases which give rise to splay, bend and twist geometries [20].

3.5.1.1 Twist geometry

An electric field is applied perpendicular to the x-axis. Let this be the y-axis. If the anisotropy of the dielectric susceptibility is positive, then the director tends to align along the direction of electric field, rotating away from the x-axis towards the y-axis. Let us call the angle between the director and the x-axis be θ. If we consider the dimensions of the flat pieces of glass to be much larger than the separation, then θ should not be a function of x or y but should depend on z (an axis normal to the surfaces of the glass). This geometry is illustrated in **Figure 14**.

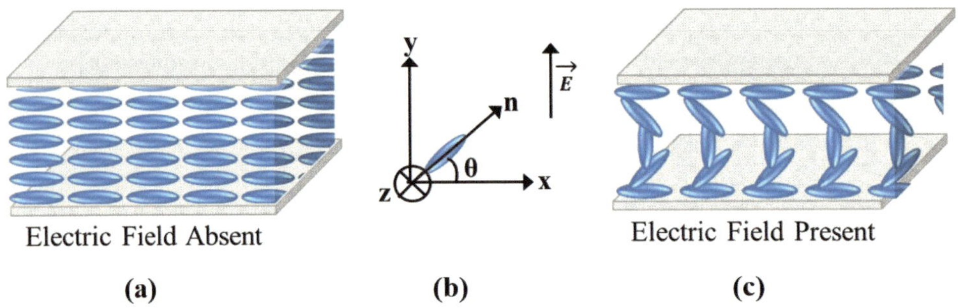

Figure 14.
Twist deformation in nematic LC molecules: (a) initially oriented planar cell, (b) sketch of molecular reorientation, and (c) deformation induced by electric field.

3.5.1.2 Splay geometry

A director is oriented along the x-axis, but now the electric field is applied in the z direction. The director now has x and z components, and $\theta(z)$ is measured from the x-axis to the director in the xz plane as shown in **Figure 15**.

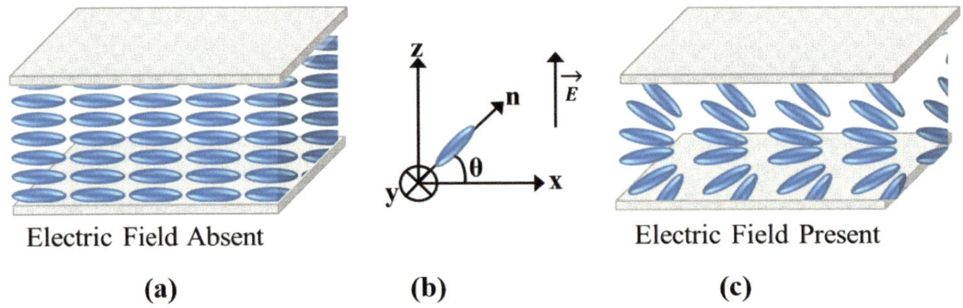

Figure 15.
Splay deformation in nematic LC molecules: (a) initially oriented planar cell, (b) sketch of molecular reorientation, and (c) deformation induced by electric field.

3.5.1.3 Bend geometry

The last geometry also involves both splay and bend. As shown in **Figure 16**, the boundary conditions are such that the undistorted director points along the z-axis and the electric field is applied along the x-axis. The angle $\theta(z)$ now is measured from the z-axis to the director in the xz plane.

The threshold value for deformations of the director $\hat{n}$ in the electric (E_t) and magnetic field (B_t) is given by

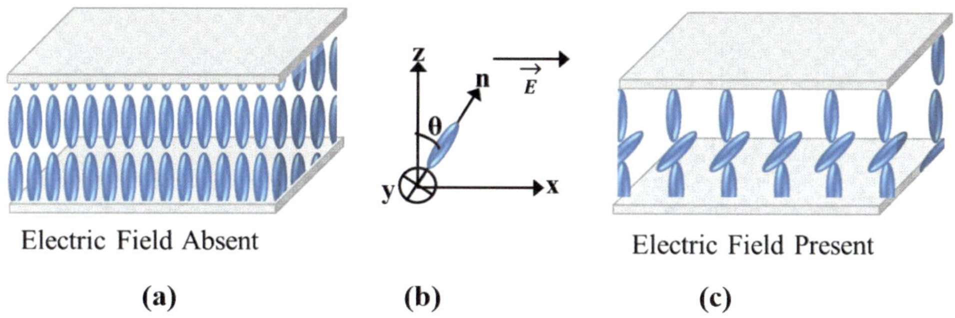

Figure 16.
Bend deformation in nematic LC molecules: (a) initially oriented homeotropic cell, (b) sketch of molecular reorientation, (c) deformation induced by electric field.

$$E_t = \frac{\pi}{d}\sqrt{\frac{K_{ii}}{\varepsilon_0 \Delta\varepsilon}} \tag{9}$$

$$B_t = \frac{\pi}{d}\sqrt{\frac{K_{ii}}{\mu_0^{-1}\Delta\chi}} \tag{10}$$

where K_{ii} is elastic constant; ii = 11, 22, 33 corresponds to splay, bend and twist deformations respectively; μ_0 is permeability of free space and $\Delta\chi$ is anisotropic diamagnetic susceptibility.

3.6 Alignment of liquid crystals

To manufacture LC device with desired electro-optic (EO) effect, confinement and alignment of LC molecules in a specific direction is very essential. Mauguin reported that LC domains could be aligned by placing them in contact with a crystal surface. The structure of LC nearby interface is different from that in the bulk. The interfacial LC molecules change the boundary conditions and influence the LC in bulk. By controlling the LC directors at the surface, reproducible director orientations can be obtained. The different interaction (anchoring) conditions of LC molecules with their neighboring phase (solid substrate) give rise to different types of liquid crystal display (LCD) devices with varied properties [4, 31, 38, 43–46]. Various types of LC molecule alignment can be induced by treating the supporting substrate differently. The most common types of alignment are homogeneous (planar) and homeotropic.

3.6.1 Homogeneous alignment

This is also called as planar alignment (**Figure 17(a)**). Here, directors of LC molecules are oriented parallel to the electrode surface. Homogeneous alignment refers to the unidirectional orientation of the molecular axis in the planar mode and displays birefringence with excellent optical quality [47]. It can be achieved using surface treatment methods, such as obliquely evaporated SiOx layers, Langmuir–Blodgett films, photoalignment and rubbed polymer films [48–50]. Out of which photoalignment and mechanical rubbing are more promising techniques. In photoalignment, materials like polyvinyl alcohol (PVA) or polyvinyl cinnamate (PVC) are coated on indium tin oxide (ITO)-coated glass plates. These materials are illuminated with polarized ultraviolet light, which forces the LC directors to align parallel to the specific surface direction. Another method is rubbing, invented by Mauguin in 1911; in this method electrode is coated with transparent polymeric

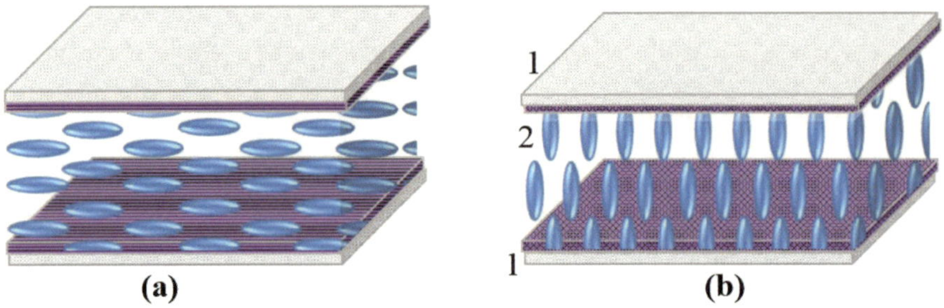

Figure 17.
(a) Homogeneous and (b) homeotropic alignment of liquid crystals. [1, PI coated ITO glass plate; 2, LC molecules].

material (generally PI), followed by baking and rubbing [45, 51]. The thin layer of PI is known for its exceptionally strong and outstanding heat, mechanical and chemical resistivity [52]. The mechanical treatment such as unidirectional rubbing modifies surface topography by breaking the symmetry and creating linear micro-grooves on the polymer surface [48, 53, 54]. The rubbing direction on one ITO plate is 0° or 90° with respect to other depending upon the parallel/antiparallel or twisted mode, respectively [55, 56].

3.6.2 Homeotropic alignment

In homeotropic alignment the LC molecule directors are oriented perpendicular to the electrode surface (**Figure 17(b)**). It can be achieved by treating the surface with a surfactant such as hexadecyltrimethylammonium bromide (HTAB), lecithin and polymers [57]. The polar head of a surfactant chemically attaches to the substrate, and the hydrocarbon tail points out, perpendicular to its surface. At this point maximum intermolecular interaction between LC molecules and surfactant promotes perpendicular anchoring of the nematic LC director. However, surfactant-based homeotropic alignment is not stable against humidity and heat [58, 59].

Apart from these two standard alignments, there are many other variations such as hybrid, twisted, supertwisted, fingerprint, multidomain vertically aligned, etc. which are employed in various LCD devices.

4. Applications of liquid crystals

LC science and applications now permeate almost all segments of the society from display technology to beyond display front. LCD is a well-known and leading application of LC [60] in the information display industry. They are used in small-sized displays such as smart phones, calculators, wearable displays and digital cameras, medium-sized displays such as desktop and laptop computers and large-sized displays such as data projectors and direct view TVs. They have the advantages of having high brightness and high resolution and being flat paneled, energy saving, light weight and even flexible in some cases [61]. To select the appropriate LCD for application and to tailor their optical performance, we need to understand broad classification (**Figure 18**) of LCD and their basic mechanism [62]. Till date mainly three types of LCD have been developed: transmissive, reflective and trans-reflective.

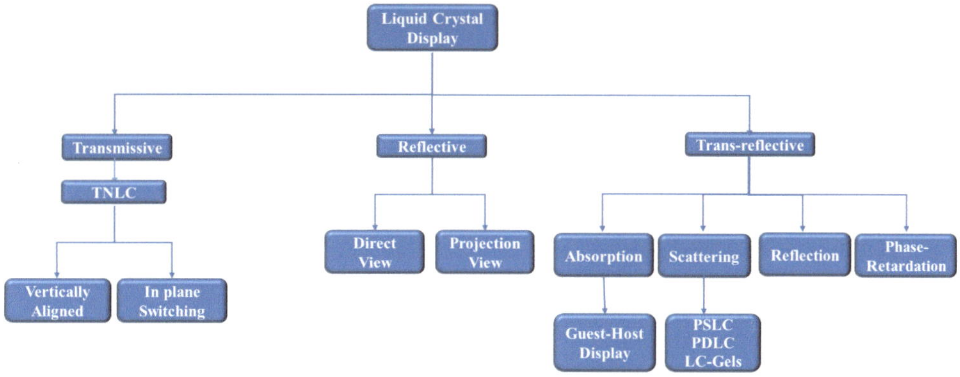

Figure 18.
Classification of LCD.

4.1 Transmissive LCD

A transmissive LCD transmits a backlight for illuminating the LCD panel, which results in high contrast ratio and high brightness. As their viewing angle is limited, they are more suitable for single-viewer applications, such as games and notebook computers. To make them applicable for multiple viewers, such as televisions and desktop computers, a phase compensation film should be introduced in them. They can also be used for projection displays, for which a high-power arc lamp or a light-emitting diode (LED) array is used as a light source. The most common and finest example of transmissive LCD is twisted nematic liquid crystal (TNLC) cells which are extensively used for notebook computers, where viewing angle is not critical. Its operating principle is based on the ability of the nematic LC to rotate the polarization of light beams passing through it [43, 63].

4.1.1 Twisted nematic liquid crystal cell

It was first invented by Schadt and Helfrich and demonstrated by Fergason in 1971 [64, 65]. It consists of two ITO-coated glass substrates, additionally coated with transparent alignment layers, usually PI. These PI-coated glass plates are rubbed with velvet cloth in one direction; as a result, the LC molecules orient parallel to the rubbing direction. The rubbing directions on two substrates are perpendicular to each other. These glass plates are arranged in such a way that a 90° twist of director from one substrate to the other is formed inside the cell. The cell is kept in between two crossed polarisers in such a way that their polarization is parallel to the rubbing direction of the same glass substrate. In the absence of electric field, the top LC alignment is parallel to the optical axis of the top polarizer, while the bottom LC directors are rotated 90° and parallel to the optical axis of the bottom polarizer (analyzer) as shown in **Figure 19(a)**. When $d\Delta n \gg 0.5\lambda$ (the Gooch-Tarry's first minimum condition) is satisfied, the incoming linearly polarized light will follow closely the molecular twist and transmit the crossed analyzer. Here Δn is the birefringence of LC, d is the cell gap, and λ is the wavelength of the light. This is called the normally white (NW) mode, since light is transmitted without application of any voltage. In the voltage-on state (**Figure 19(b)**), the LC molecules undergo a Freedericksz transition. In this state, the director of the nematic LC is parallel to the field and no longer twisted. When polarized light enters a cell in such a configuration, it is not twisted and is absorbed/blocked by the analyser, resulting in a dark state. Regions where an electric field is applied appear dark against a bright

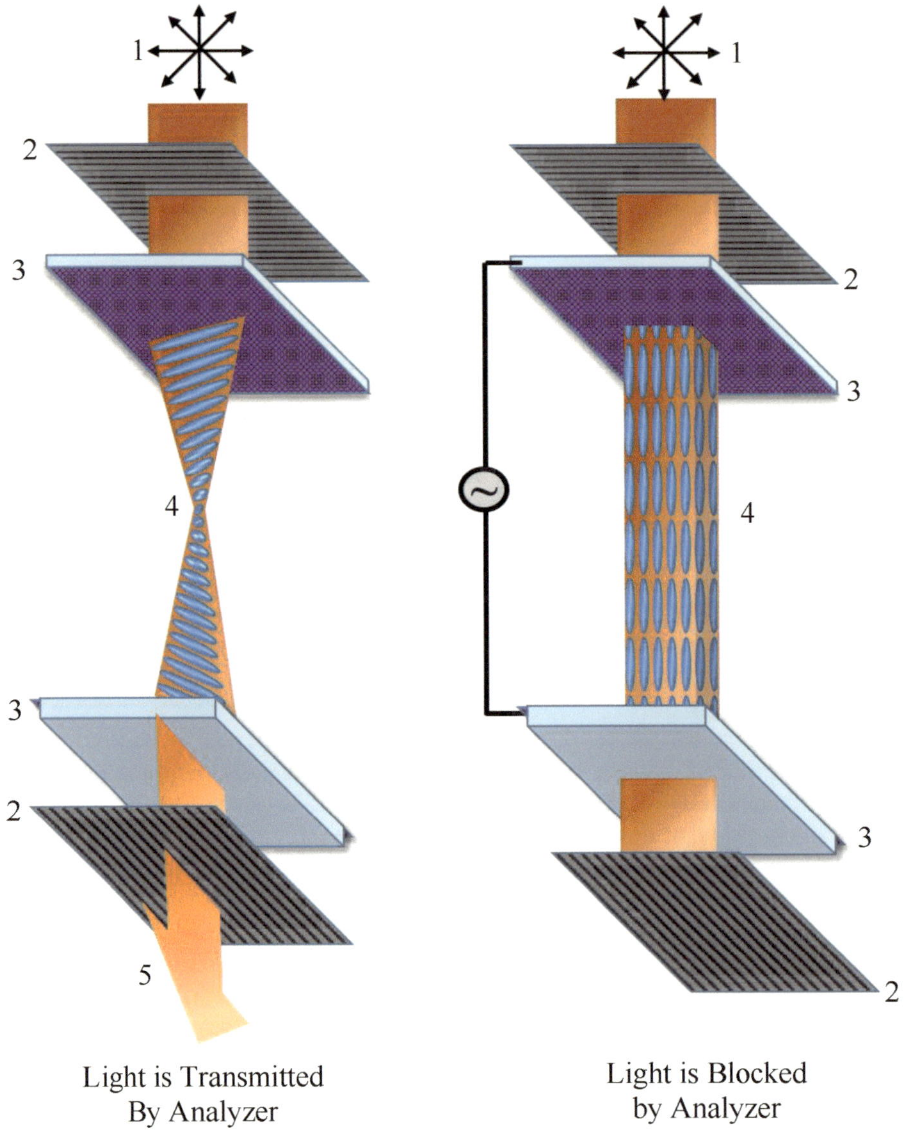

Figure 19.
Twisted nematic LCD in (a) OFF state and (b) ON state. [1, unpolarised light; 2, polarizer; 3, PI coated ITO glass plates; 4, LC droplets; 5, polarized light].

background. Because of the orthogonality of boundary layers, the dark state is achieved at relatively lower voltage. Depending on the field strength, twisted nematic displays can switch between light and dark states, or somewhere in between (greyscale.) How the LC molecules respond to applied field is the important characteristic of this type of display. However, every device has some shortcoming, in TNLC is its narrow viewing angle and poor color production. To overcome these problems, new technologies such as in-plane switching and vertical alignment mode have been introduced [61].

4.2 Reflective LCD

In R-LCDs, the necessity of backlight source (as in transmissive-type LCD) has been seized. They reflect ambient light for displaying images. Therefore, they consume low power, are lighter in weight and have good readability in outdoor environment, but are inapplicable under low or dark ambient conditions. The R-LCDs are of two types: direct view and projection view.

4.3 Trans-reflective LCD

In order to overcome the drawbacks and to take advantage of both transmissive as well as reflective LCDs, trans-reflective LCDs have been developed, which use both ambient light and backlight to display images based on availability and necessity. It has a semi-reflective film in the back of LCD screen, the backlight can transmit through it so that it may work as a transmissive mode, but the front light cannot pass through it and get reflected, and it simultaneously works as a reflective mode. Broadly, trans-reflective LCDs are classified into four categories: (a) absorption, (b) scattering, (c) reflection and (d) phase retardation. As the name suggests, the first category absorbs light, and the corresponding device is referred to as guest-host (GH) display. The second one scatters light, and polymer-dispersed liquid crystal (PDLC), polymer-stabilized liquid crystal (PSLC) and LC gels are related technologies. The third category is based on reflection of light. The fourth one modulates the phase of an incident light. Here we will discuss operating principle of PSLC, PDLC and holographic PDLC (HPDLC).

4.3.1 Guest-host display

These types of display systems were first introduced by Heilmeier and Zanoni [66]. As the name suggests, in these systems light is absorbed by the guest material, which are generally dichroic dye molecules, dissolved in a host LC material. Dichroic dye molecules are rod-shaped molecules, which absorb light of certain wavelength more along one axis than the other. Dye molecules get dissolved in LC and orient along with LC molecule. Upon application of external field, the rotation of absorption axis of dye molecule along with LC molecule modulates light transmission. Mainly, there can be three types of GH displays exist: the Heimeier type, the double-layer type and the PDLC type, which uses 1, 2 and zero polarizers, respectively [15, 67].

4.3.2 Polymer-liquid crystal composites

Polymer-LC composites are thin films prepared from phase-separated high molecular weight polymer and low molecular weight LCs. The polymer, which is homogenously mixed into LC, provides mechanical and structural stabilization to LC devices. Polymer-LC composites have been used in a wide range of applications such as high-definition spatial light modulators; switchable windows; flat-panel large area flexible displays; light valves; color projectors; thermal, optical and strain sensors; bi-stable reflective displays and haze-free normal- and reverse-mode light shutter devices [68–71]. The confinement of LC material in both dense polymer matrix and moderate polymer networks modifies the bulk LC phase. Depending upon the concentration of monomer unit, they are classified as polymer-stabilized liquid crystal or polymer-dispersed liquid crystal or holographic polymer-dispersed liquid crystal (HPDLC). In PSLC, polymer forms a sponge-like structure in a continuous LC medium. The concentration of LC is much higher than the polymer concentration. In PDLC, the LC is in the form of micron- and submicron-sized droplets, which are dispersed in a continuous polymer matrix. The concentration of polymer is comparable to the LC [61]. In HPDLC the polymer concentration is high around 60–70 wt%. As droplet size is much smaller than the visible wavelength,

HPDLC films are free of light scattering. These films have faster response time and require higher switching voltages. In PDLC and HPDLC films, no surface alignment layer is needed [72].

5. Polymer-stabilized liquid crystal

Polymer-stabilized liquid crystal is a thin composite film prepared from the homogenous mixture of LC and monomer. Typically, the monomer concentration in PSLC is less than 10% of the total weight. The small amount of monomer is used to stabilize/lock the oriented LC structure at different optical states and to reduce the switching time and operating voltage [73]. The homogenous mixture of LC and polymer is prepared and filled into the prefabricated cells made of transparent ITO-coated electrodes for photopolymerization using PIPS technique [51]. For improved display performance, LCs should be homogenously and uniformly aligned inside the cell. To control the orientation of LC, the inner face of electrodes is coated with a transparent polymeric material (generally PI) followed by baking and rubbing [53]. A thin layer of PI is known for its excellent strong and outstanding heat, mechanical, and chemical resistivity [53]. The mechanical treatment such as unidirectional rubbing modifies surface topography by breaking the symmetry and creating linear microgrooves on the polymer surface [48, 54, 56]. The rubbing direction on one ITO plate is 0° or 90° with respect to the other depending upon the parallel/antiparallel or twisted mode, respectively [55, 74]. This induced anisotropic surface diffuses monomer molecules preferentially along the rubbing direction. Due to strong interaction and anchoring between LC and monomer, the polymer network formed during/after polymerization keeps the LC director in a definite direction [75]. Along with the surface alignment layer, the configuration and orientation of LC can also be controlled by application of external field and/or temperature during photopolymerization. Even low electric field is sufficient to align the LC director along the field by fixing torque on it. After establishing the proper combination of surface treatment and applied field, the sample is irradiated by ultraviolet (UV) light to induce photopolymerization to obtain the desired texture. Since the monomer concentration is very small, a continuous LC texture along with interconnected, interpenetrating and mitigated polymer network can be obtained after polymerization [61, 76–78] as shown in **Figure 20**.

Because of application of electric field during photopolymerization, oriented LC domains are formed. Also, this controlled alignment of the LC molecules between polymer networks has significant effect on the transmittance, absorbance, response time, and dielectric properties of PSLC films [79].

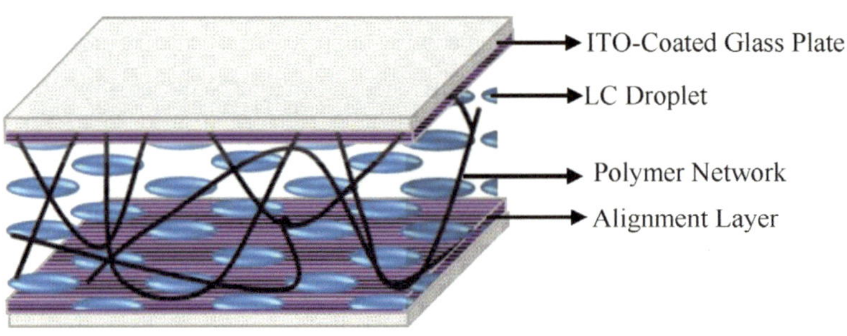

Figure 20.
Polymer stabilized liquid crystal.

5.1 Morphological analysis

Figure 21 shows POM images of PSLC films which are prepared using LC BL036 and prepolymer NOA-65 in 95/5 wt/wt% ratio under different rubbing directions and in the absence and presence of electric field [80].

Figure 21 shows the POM image of four types of PSLC films named as "A" (antiparallel rubbing; electric field absent), "B" (antiparallel rubbing; electric field present), "C" (twisted rubbing; electric field absent) and "D" (twisted rubbing; electric field present). On the acute observation of these images in all the four samples, complex geometrical structures of LC and polymer network were found. Samples A and C which were prepared without applying any electric field during polymerization showed rectilinear alignment. Since polymer network has a strong aligning effect on the LC, therefore it tends to keep LC in the aligned state [81]. In these films, polymer chains move throughout the sample parallel to the rubbing direction; therefore shorter chains with smaller domains entrapped between them were observed. In the case of samples B and D, which were prepared by applying electric field during polymerization, bigger LC domains were formed. Upon application of electric field during sample preparation, LC material orients and often indefinitely retains the alignment imposed by an electric field [82]. Because of the adsorption of LC droplets at the polymer wall, polymer network grows in the direction of field. Due to which cross linked, thicker and topologically defective polymeric walls were formed. Also, diffraction of light was observed from polymer nodules.

The effect of electric field on the orientation and confirmation of the LC material between the boundaries of polymer network gives deep insight in understanding PSLC film behaviour.

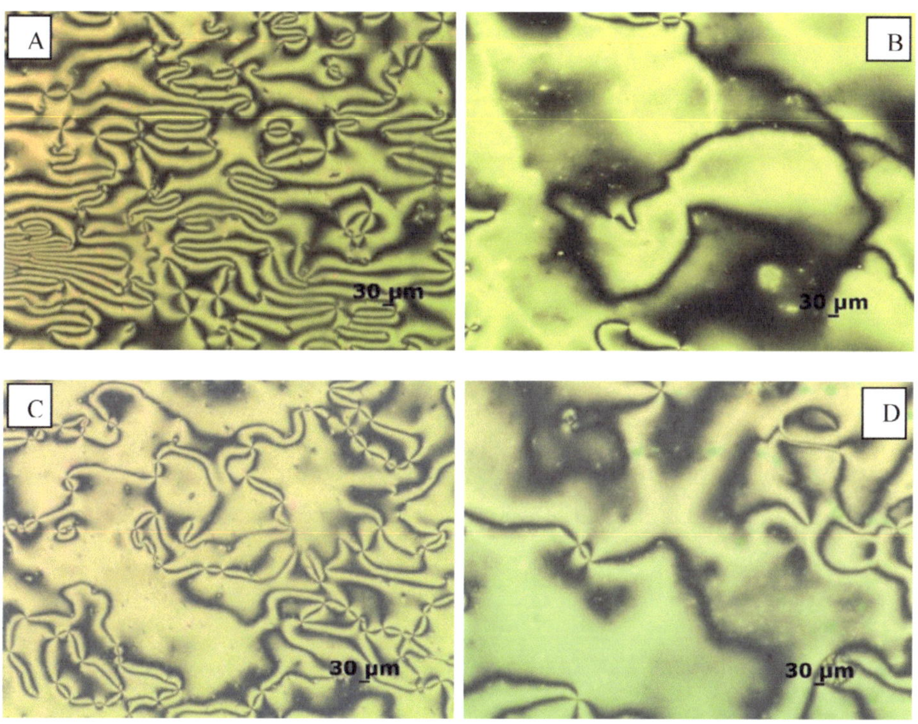

Figure 21.
POM images of homogenously aligned PSLC films: (A) antiparallel rubbing cured without any voltage, (B) antiparallel rubbing cured by applying 10 V, (C) 90° twist rubbing cured without any voltage and (D) 90° twist rubbing cured by applying 10 V, observed using 5× objective.

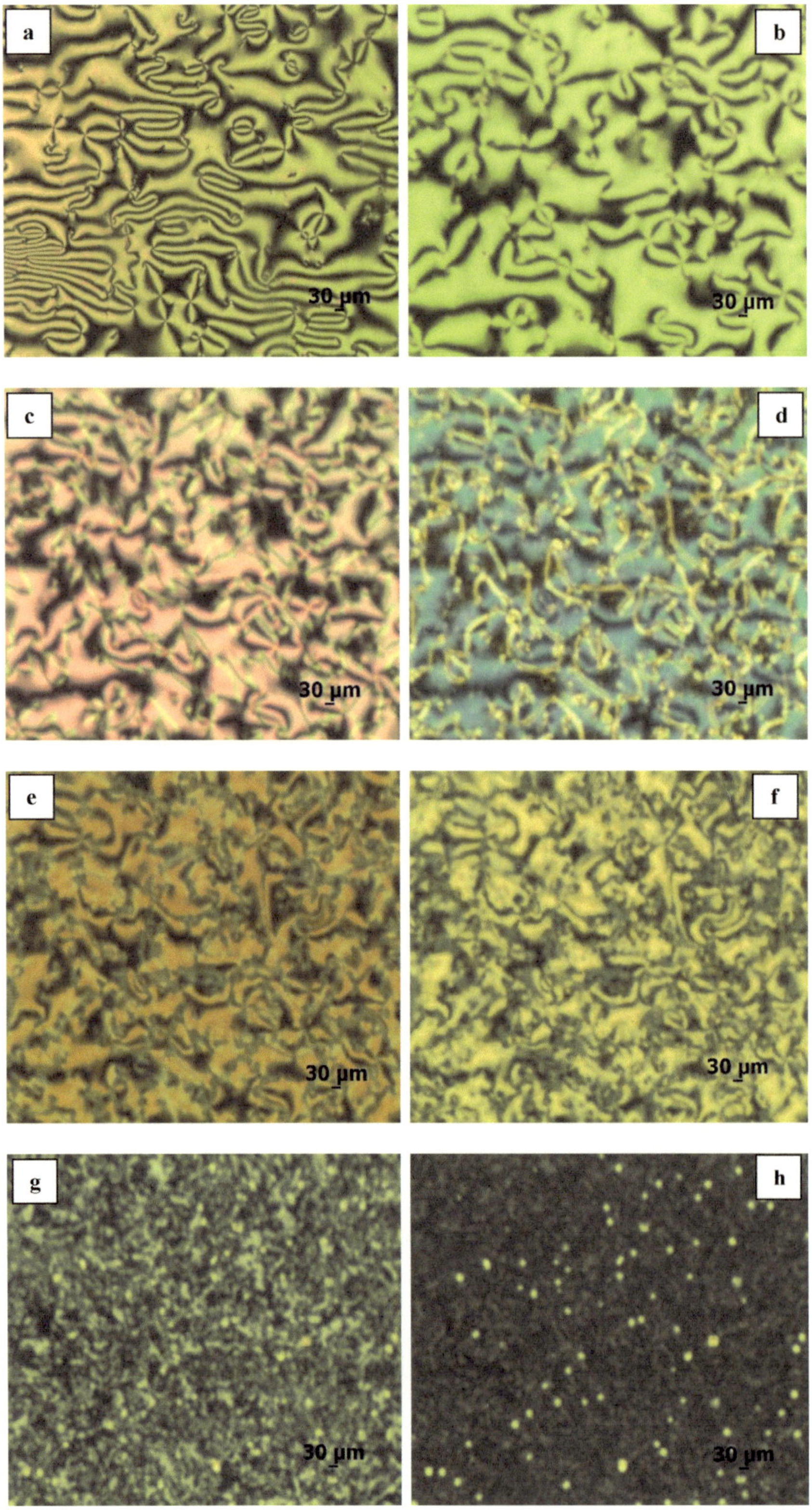

Figure 22.
Effect of voltage on homogenously aligned PSLC film sample A (anti-parallel rubbing; electric field absent during polymerization) observed at (a) 0 V, (b) 2 V, (c) 6 V, (d) 10 V, (e) 20 V (f) 40 V, (g) 60 V and (h) 80 V using 5× objective lens.

Figure 22 shows effect of electric field on sample A, prepared with antiparallel rubbing direction and in the absence of field. Upon application of low electric field of ~10 V, change in color was observed, which indicates change in refractive index. The change in refractive index is basically due to orientation of LC molecules in the domains. Also ~10 V vivid polymeric boundaries were made visible due to adsorption of LC molecules on frozen polymer network. On the application of high electric field of ~40 V and above, dark homeotropic state with the LC directors aligned perpendicular to the substrate surface was observed [83].

5.2 Voltage dependence of transmittance at fixed frequency

The voltage-transmittance curve (**Figure 23**) of PSLC film indicates that it follows an opposite trend to that of the PDLC films. It is clear from the graph that PSLC film has maximum transmittance (T_{OFF}) when no voltage is applied. Upon application of electric field, the LC orientation changes from planar to homeotropic state of alignment, and the transmission decreases rapidly for PSLC film.

At a particular voltage, transmittance of PSLC decreases by additional 10% of the OFF-state transmittance; this voltage is termed as threshold voltage (V_{TH}). With the further increase in voltage, the polarization state of light is perpendicular to the analyser which gives dark state or minimum transmittance (T_{ON}) state. A definite value of voltage at which PSLC film achieve T_{OFF} state is termed as ON-state voltage (V_{ON}) condition.

The ratio of maximum to minimum transmittance gives contrast ratio (CR) of the film, and difference between maximum and minimum transmittance gives transmittance difference (ΔT). Mathematically

$$CR = \frac{T_{OFF}(\%)}{T_{ON}(\%)} \tag{11}$$

$$\Delta T(\%) = T_{OFF}(\%) - T_{ON}(\%) \tag{12}$$

Table 2 gives voltage-transmittance data of PSLC film prepared using monomer NOA-65 and LC BL036 in 95/5 wt/wt%.

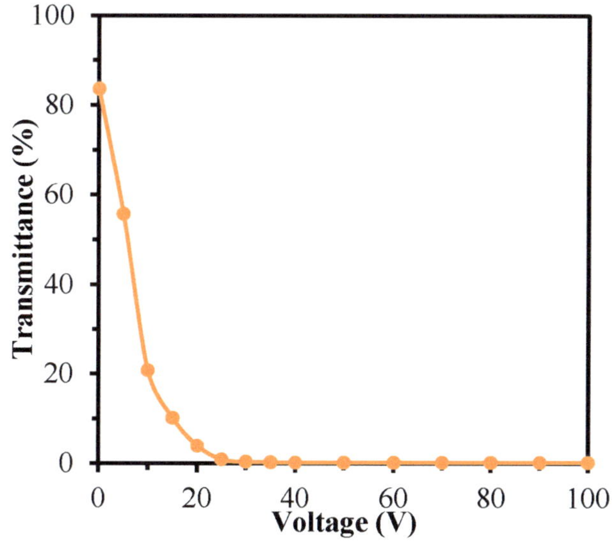

Figure 23.
Transmittance vs. voltage curve of PSLC film.

Sample	T_{OFF} (%)	T_{ON} (%)	ΔT (%)	CR	V_{TH} (V)	V_{ON} (V)
N36	83.6	0.20	83.4	418	2	25

Table 2.
Voltage-transmittance data of PSLC film prepared using monomer NOA-65 and LC BL036 in 95/5 wt/wt% ratio.

5.3 Hysteresis effect

In a scan up cycle, as voltage increases from 0 V to V_{max}, transmittance of PSLC films decreases from T_{OFF} to T_{ON}, whereas in scan down cycle as voltage decreases from V_{ON} to 0 V, transmittance increases from T_{ON} to T_{OFF}, but it does not follow the same path. The above observed phenomenon is termed as hysteresis and should be minimized for better electro-optic properties. It was observed that at a given voltage, transmittance for scan up cycle was higher than the transmittance for scan down cycle. A measure of hysteresis is given by the voltage width at half of maximum transmittance (ΔV_{50}). The hysteresis behaviour of a PSLC composite film is shown in **Figure 24**. Also, hysteresis effect was not observed at high fields because at higher voltages of scan down cycle, LC domains remain in the same state of orientation. However, when the applied field is reduced further, reorientation of LC domains begins, giving rise to hysteresis effect.

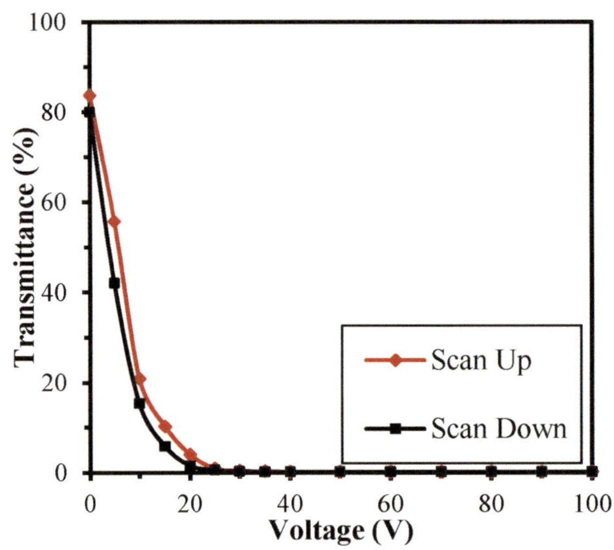

Figure 24.
Hysteresis curve of PSLC film.

5.4 Response time

Response time is the sum of rise time (τ_r) and decay time (τ_d). Upon application of electric field, LC molecules align along the field, and transmittance of film decreases in PSLC. The time in which transmittance of film reaches from 90–10% is termed as rise time. On the removal of field, these LC molecules relax back to their initial position. The time in which transmittance reaches from 10–90% is termed as decay time. Variation in rise and decay time of PSLC films with respect to applied voltage is shown in **Figure 25**. However, with the increase in voltage rise time shortens, but the trade-off is longer decay time [83, 84].

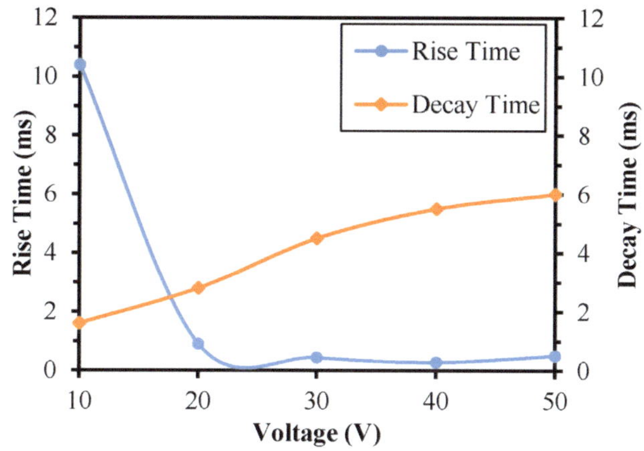

Figure 25.
Rise and decay time vs. voltage curve of PSLC film.

5.5 Dielectric properties of polymer-LC composite films

Polymer-LC composite films are complex heterogeneous system, holding intrinsic anisotropy of LC and polymer. In order to gather the information about the structure, alignment, phase transitions and intermolecular interactions of composite films, knowledge of their dielectric properties is essential [85, 86]. For this purpose dielectric relaxation spectroscopy (DRS) is one of the best methods to measure the dielectric constant and associated parameters with high accuracy and sensitivity in polymer-LC composites. It is based on a concept of "energy storage" and resulting "relaxation" per release of this energy by the system's individual components. By developing analogy between polymer-LC composites and passive electrical circuit, polymer-LC composite films can be conveniently illustrated as a parallel plate capacitor. Here, two ITO-coated glass plates act as a parallel electrode with plate separation d and plate area A, and polymer-LC material acts as a dielectric material as shown in **Figure 26(a)**. The effective circuit of polymer-LC cell is shown below (**Figure 26(b)**), where R_o is the resistance of electrodes and R_{LP} and C_{LP} are resistance and capacitance of polymer-LC layer, respectively.

The capacitance is measured, and the relative permittivity ε_r can be calculated from the formula given below:

$$C_p = \frac{\varepsilon_0 \varepsilon_r A}{d} \tag{13}$$

where $\varepsilon_0 = 8.854 \times 10^{-12}$ F/m is the permittivity of free space. Relative permittivity ε_r is complex quantity, also known as dielectric constant; it depends on parameters like temperature, pressure and frequency. To elucidate the frequency dependence of ε_r, we must understand the different polarization mechanisms that contribute to the dielectric constant (permittivity). In response to an applied electric field, various types of polarisations may arise, such as electronic, ionic, orientational and interfacial [87].

a. Electronic/atomic polarization: Electronic polarization occurs when the electric field displaces the centre of a negatively charged electron cloud relative to the positive nucleus of the atom and induces a dipole moment. It has been found to be independent of frequency and vanishes as soon as the electric field is removed [88, 89].

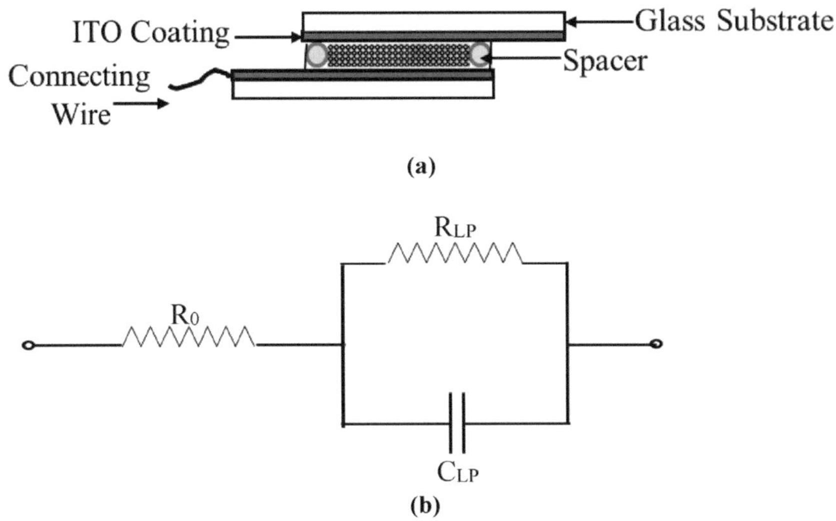

Figure 26.
(a) Schematic of polymer-LC composite film and (b) polymer-LC composite films as RC-circuit.

b. Ionic/displacement polarization: Ionic polarization also called as vibrational polarization occurs in ionic substances and is related to the displacement of atoms, causing the separation of charges [90, 91].

c. Dipolar/orientational polarization: Dipolar polarization usually occurs in materials with permanent dipoles. Under normal conditions, these materials exhibit zero net dipole moment and polarization, as the dipoles in these materials are randomly distributed/oriented. Upon application of an external electric field, the dipoles tend to orient along the direction of applied field, resulting a non-zero net dipole moment and polarization [90, 91].

d. Interfacial/translational polarization: Interfacial polarization is associated with migrating charges, by electrons or ions, over macroscopic distances in an applied field. These charges get trapped and accumulate at physical barriers such as defects, impurities, voids and grain or phase boundaries. The accumulation of charges distorts the local electric field and causes permittivity. Interfacial polarization is most prevalent in heterogeneous system like polymer-LC composites and is usually observed at lower frequencies. In general, a given dielectric material exhibits more than one polarization mechanism, and the average dipole moment in a given material is the sum of all polarization contributions. All four types of polarization occur at low frequency. Each of the above-mentioned polarization processes has specific features in the frequency and temperature dependence of the real and imaginary part of the complex dielectric permittivity. **Figure 27** shows the frequency dependence of various types of polarisations [90–92].

As frequency increases the contribution from each type of polarization successively decreases because polarization can no longer follow the variation in the field. As a result, there is a decrease in dielectric constant (relative permittivity) with the increasing frequency. The frequency dependence of dielectric constant reflects the fact that a material's polarization does not respond instantaneously to an applied field. The response must always arise after the applied field which can be

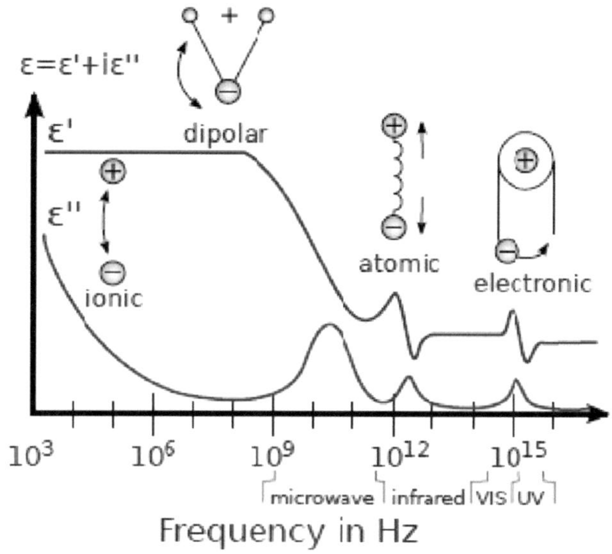

Figure 27.
A Dielectric constant Spectrum over a wide range of frequencies. ε′(ω) and ε′′(ω) denote real and imaginary part of the dielectric constant (relative permittivity) respectively.

represented by a phase difference. For this reason, dielectric constant is often treated as a complex function of the angular frequency (ω) of the applied field [93, 94].

5.5.1 Complex dielectric constant

The frequency-dependent permittivity characterizes amplitude and timescale (via the relaxation time) of the charge-density fluctuations within the sample. For the evaluation of relaxation time (time required for LC droplet reorientation), dielectric permittivity is expressed as a complex function of angular frequency (ω) of applied field:

$$\varepsilon^*(\omega) = \varepsilon'(\omega) - i\varepsilon''(\omega) \tag{14}$$

where $\varepsilon'(\omega)$ and $\varepsilon''(\omega)$ are real and imaginary parts of the complex dielectric constant.

Relaxation processes are characterized by a step-like decrease of the real part ε' and a peak in the imaginary part $\varepsilon''$of the complex dielectric function $\varepsilon^*(\omega)$ with increasing frequency. The real part is related to stored energy also called as dispersion, whereas imaginary part is related to loss of energy or dissipation called as absorption of the system. It is reasonable to introduce here a quantity "loss tangent", which is a measure of the energy dissipated due to oscillating field also known by dissipation factor "D" [21, 92, 95]:

$$tan\delta = \frac{\varepsilon''(\omega)}{\varepsilon'(\omega)} \tag{15}$$

For parallel plate capacitors with ideal dielectrics, the loss angle δ can be graphically expressed as shown in **Figure 28**.

Here, V is the applied voltage and I_c and I_R are the vector components of current I. The current I_c represents a non-lossy capacitive current proportional to the charge stored in the capacitor; it is frequency dependent and leads voltage by 90°. The current I_R is the alternating conduction current in phase with the applied voltage V, which represents the energy loss or power dissipated in the dielectric [96]. If ψ is the phase difference between the potential and current, then

$$\delta = 90 - \psi \tag{16}$$

To promote maximum energy storage in a capacitor, the dielectric loss, originating from interfacial, dipolar, distortional and conduction losses, should be minimal [87]. In general, dielectric loss increases with increase in humidity, temperature, frequency and amplitude of the applied voltage for most of the materials.

5.5.2 Macroscopic models for dielectric spectra

Dielectric relaxation processes are usually analyzed using model functions. Starting from the Debye function, several formulas for both the frequency and the time domain have been suggested to describe the experimentally observed spectra. The most important approaches are discussed below [97, 98].

a. Debye model: It is a method to study the dielectric behaviour of a material by measuring the complex dielectric permittivity versus frequency at constant temperature and ambient pressure. As the dielectric spectrum is obtained in a frequency domain, it is called as a frequency domain dielectric spectroscopy (FDDS). If a single relaxation is considered, it is known as Debye-type relaxation, and the time assumed for it is Debye relaxation time which is inversely related to the critical relaxation frequency. It is the point where dissipation factor is maximum. The relaxation frequency f is related to relaxation time τ by the relation

$$\tau = \frac{1}{\omega} = \frac{1}{2\pi f} \tag{17}$$

The dispersion and absorption terms for single relaxation as a function of the field angular frequency ω and relaxation time τ are given as

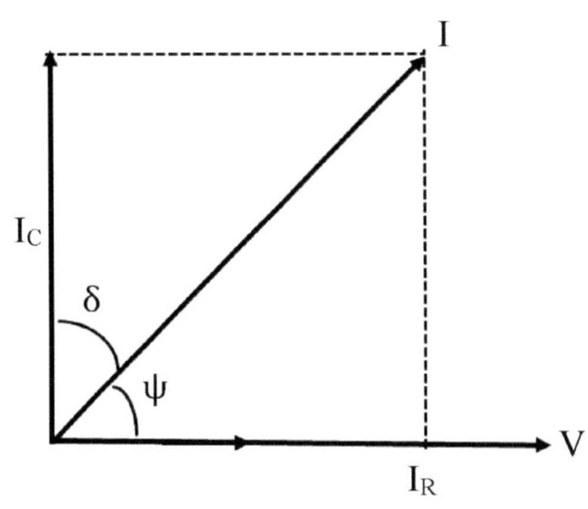

Figure 28.
Graphical representation of loss tangent.

$$\varepsilon'(\omega) = \varepsilon_\infty + \frac{(\delta\varepsilon')}{1+\omega^2\tau^2} \tag{18}$$

$$\varepsilon''(\omega) = \frac{(\delta\varepsilon')\omega\tau}{1+\omega^2\tau^2} \tag{19}$$

Debye relaxation is usually expressed in terms of the complex dielectric constant $\varepsilon^*(\omega)$ of a medium. On putting values from Eqs. (18) and (19) in (14), we get:

$$\varepsilon^*(\omega) = \varepsilon_\infty + \frac{(\delta\varepsilon')}{1+(i\omega\tau)} \tag{20}$$

where $\delta\varepsilon' = \varepsilon_s - \varepsilon_\infty$ is dielectric strength of the material, which is the voltage a material can withstand before breakdown occurs. ε_s and ε_∞ are static (at 20 Hz frequency) and optical (at relaxation frequency *f*) values of the relative dielectric constant, respectively, which were obtained by experimental relaxation spectra [91, 99–101]. The frequency dependence of real and imaginary components of complex dielectric constant of PSLC film is shown in **Figure 29(a)** and **(b)**, respectively.

b. Cole-Cole model: To describe secondary relaxations, Cole-Cole model has been used, given by the equation

$$\varepsilon^*(\omega) = \varepsilon'(\omega) - i\varepsilon''(\omega) = \varepsilon_\infty + \frac{(\delta\varepsilon')}{1+(i\omega\tau)^{1-\alpha}} \tag{21}$$

where α is known as the distribution parameter and other terms are the same as in Debye model. The exponent α characterizes the breadth of the relaxation time distribution and ranges from 0 (infinitely broad distribution) to 1 (Debye's single relaxation time limit), describing different spectral shapes. When α = 0, the Cole-Cole model reduces to the Debye model. The graph drawn between imaginary part ε″ and the real part ε′ of the dielectric constant with frequency as a parameter, shown in **Figure 30**, is known as a Cole-Cole plot [102, 103]. It is useful for the interpretation of molecular dynamics of materials which possess one or more well-separated relaxation processes with comparable magnitudes. How well the ε′ and ε″ are fitted to form semicircle is an indication of the nature of relaxation behaviour. All the above-mentioned parameters were determined from fitting the experimental data of dielectric spectra with

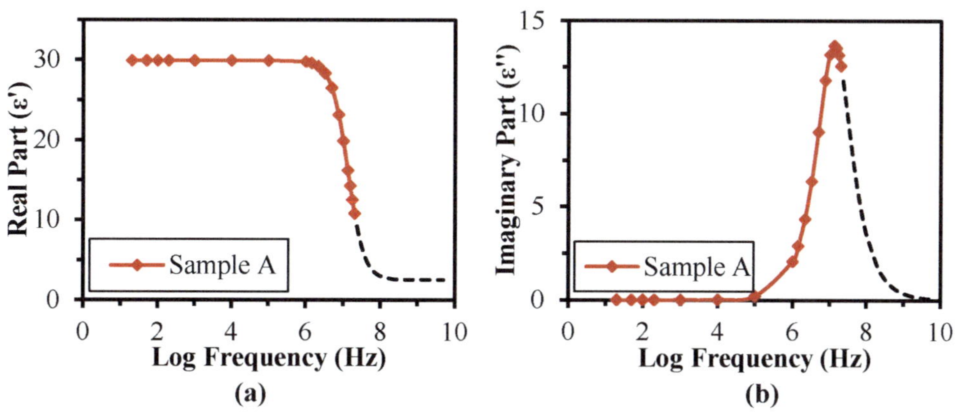

Figure 29.
Frequency dependence of (a) real component and (b) imaginary component of complex dielectric constant of PSLC film.

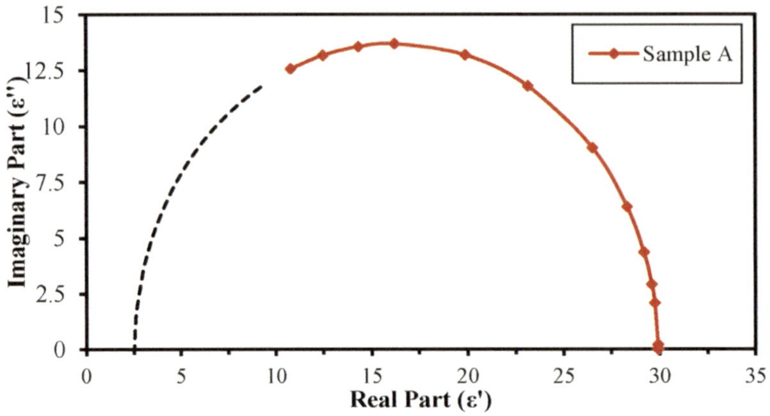

Figure 30.
Cole-Cole plot of PSLC film at 25°C temperature.

Sample	f (MHz)	ε_s	ε_∞	$\delta\varepsilon'$	τ (s)	α
A	13.2	29.9	2.53	27.4	1.21 E − 08	0.022

Table 3.
Fitting parameters of various PSLC films.

Cole-Cole approach of the Debye equation and are shown in **Table 3**. The frequency corresponding to the top point of this semicircle curve is the relaxation frequency f of orientational polarization of LC domains. At this point dielectric heating is maximum due to which dissipation factor is also maximum. The angle φ between arc radius and ε′-axis gives distribution parameter α:

$$\varphi = \frac{\alpha\pi}{2} \tag{22}$$

If the centre of the semicircle lies on the ε′-axis, then the distribution parameter $\alpha = 0$ (Debye type), and if the centre is below the ε′-axis, then $\alpha \neq 0$ (non-Debye type), while if $\alpha > 0.5$, there could be more than one relaxation process. The calculated value of α indicates that the PSLC film exhibits non-Debye-type relaxation process [104–106].

5.6 Conclusions of PSLC study

PSLC is a reverse to the conventional PDLC but identical to the twisted nematic liquid crystal cell having maximum and minimum transmittance under crossed polarizer in the OFF and ON states, respectively. However, the threshold voltages of PSLC are much lower than TNLC [107–109]. The PSLC are useful for bi-stable reflective displays and normal- and reverse-mode light shutters [68, 110]. In order to improve electro-optic responses of PSLC devices, LC material are doped with dye and nanoparticles [111, 112].

6. Polymer-dispersed liquid crystal

Polymer-dispersed liquid crystal is a smart, inhomogeneous thin composite film of micron-sized nematic LC droplets, randomly dispersed and embedded in optically isotropic polymer matrix first introduced by Fergason in 1984 [61, 113, 114]. The composite of these two is technologically very important because it encompasses various unique properties of LCs, which are mechanically and structurally strengthened by polymer matrix. The operation of these composite films is based on birefringence property of LC. For LC with positive birefringence, $\Delta n>0$, in the OFF state as the LC droplets are randomly oriented, light repetitively refract/scatters at multiple polymer-LC interfaces. Due to which most of the light incident on PDLC device scatters, and the film becomes/appears opaque (**Figure 31(a)**). On the contrary, in the ON state as the directors in LC droplets align along the direction of electric field and if the RI, n_p of polymer matrix, matches with the ordinary RI, n_o of LC, most of the normally incident light on PDLC device behaves as an ordinary light and transmits through it, and the film becomes a transparent one [22, 115] (**Figure 31(b)**). In a similar manner, for LC with negative birefringence, $\Delta n < 0$, in the ON state extraordinary RI n_e comes into picture [34]. The LC droplet structure, their interaction with the polymer matrix and their optical and dielectric anisotropy play crucial role in shaping/modeling system properties. Although LC droplets are spherical in shape, they get deformed when embedded inside a polymer matrix. The polymer is supposed to act as a mere matrix for the LC aggregates, but their physical interactions can influence the formation of mesophases. The LC droplet morphology depends on many physical parameters such as refractive indices, conductivity, type and proportion of materials used, phase separation method, rate of diffusion, viscosity and solubility of the LC in the polymer, addition of dopant or dye molecules, etc. [116–121]. Unlike other LC technologies, PDLC do not require alignment layers and polarizers. This reduces cost; simplifies design; increases device lifetime in high-temperature and high-humidity conditions; increases transmittance, contrast ratio, flexibility and mechanical strength; and reduces response time. Also, the large surface-to-volume ratio of the composite film supports the construction of large area PDLC devices [19].

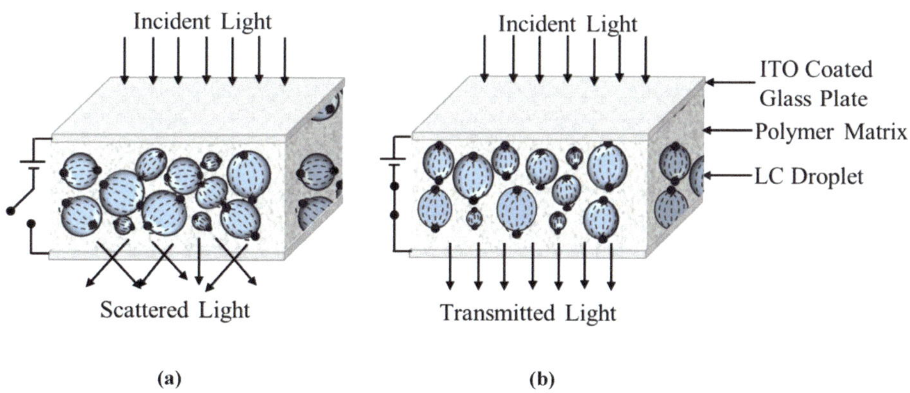

Figure 31.
PDLC in (a) OFF state and (b) ON state.

6.1 Fabrication of PDLC films

For the uniform dispersion of micron-sized nematic LC droplets inside polymer matrix, principally two methods have been enlisted, namely, phase separation and encapsulation.

6.1.1 Phase separation

The phase separation method starts with homogenous, isotropic solution of LC and polymer or prepolymer, followed by the formation of nucleated LC droplets (phase separation) and then finally solidification of polymer matrix pioneered by Doane and co-workers [122]. The morphology of a phase-separated PDLC film depends on the chemical nature of the LC and polymer constituents and kinetics of the processes driving the phase separation to occur, which ultimately control the electro-optic properties of device. Therefore, to obtain desired electro-optic properties in a device, systematically processed phase separation of LC and polymer is one of the key parameters. This technique is preferred where film moldability is required because during phase separation, polymer plasticization takes place. To induce phase separation in PDLC, generally any of the three routes, namely, polymerization-induced phase separation (PIPS), solution-induced phase separation (SIPS) and temperature-induced phase separation (TIPS), has been practised. Each method produces PDLC film with different properties and characteristics.

a. Polymerization-induced phase separation: It is usually an irreversible process, in which initially miscible, single-phased mixture of prepolymer (monomer or oligomer) and LC is filled into a prefabricated cell. On application of suitable energy in the form of heat, light or radiation, the polymerization is induced. The growing polymer chains phase separate LC droplets by forming an enclosing polymer matrix because LC is less soluble in polymer than in prepolymer. The phase-separated system has lower entropy than the mixed one. Polymerization may be initiated by heat, photo (UV irradiation) or free radicals. Generally, photopolymerization is preferred over the other two because it has low activation energy, has a good control over final properties of PDLC film and can occur in broader temperature range [120, 123, 124]. Phase separation process continues even if the UV exposure is interrupted during the cure process. Nowaday's photopolymerization technique is also applied in mask patterning.

b. Solution-induced phase separation: In this method, a polymer and LC are mixed and dissolved in a common organic solvent to form a single-phase homogeneous mixture. Such solution is then poured (using suitable technique) on an optically flat, transparent and conducting substrate for solvent evaporation. Evaporation of solvent causes phase separation between polymer and LC domains. Droplets start growing as the polymer and LC come out of solution and grow up till the polymer solidification. Another substrate (generally ITO) is laminated and annealed on the film using pressure and heat to form a PDLC cell. The electro-optic properties, which depend on LC droplet size and morphology, can be tailored by controlling evaporation rate. Droplet size and shape can be manipulated after the PDLC cell formation, by heating the sample cell till the clearing temperature of LC and then cooling. During heating LC get dissolved in polymer and again phase separates on cooling. By varying cooling rates, LC droplet size can be varied.

c. Temperature-induced phase separation: It is useful for thermoplastics which melt below their decomposition temperature. Thermoplastic and LC are heated until the clear solution is formed, such solution is filled into a prefabricated cell and then cooled. When the system is cooled, the LC phase separates from the solidifying polymer. Droplet size can be tailored by controlling the rate of cooling. Droplet size (droplet numbers) shows inverse (direct) relation with rate of cooling. Methodically, it is a simple way of phase separation, but it is less preferred because the properties of TIPS films are difficult to reproduce, often sensitive to the processing history, and homogenous mixture of LC and thermoplastic is obtained at relatively higher temperature than in PIPS and SIPS techniques, which may worsen electro-optic properties of the PDLC device.

6.1.2 Encapsulation method

In contrast to phase separation method, encapsulation method starts from inhomogeneous solution. LC is dispersed by rapid stirring as an emulsion in an aqueous solution of a film-forming polymer such as PVA. This emulsion is then spin-coated or deposited onto a transparent conductive substrate like ITO-coated glass plate and dried. Thousands of non-uniform, sometimes interconnected LC capsules (droplets), surrounded by a solid layer of polymer are produced. Size of these droplets depends on stirring speed and time. Materials manufactured from this method are also known as nematic curvilinear aligned phase (NCAP). Encapsulation method is beneficial as LC is insoluble in aqueous solution; equilibrium phase separation is easily achieved as well as polymer plasticization is prevented. But it has limitation also; because of water evaporation, there is significant change in volume of film which tends to deform droplet structure, and only few polymers are appropriate for encapsulation with proper optical properties [120, 123, 124].

6.2 Nematic configurations in PDLC films

The nematic material confined in a droplet in a PDLC is in a particular arrangement, called the director configuration. LC droplets are usually spherical because of surface tension, but due to photopolymerization reaction the texture changes significantly to adopt different configurations. When LCs are confined to small cavities, curved surfaces deform the director field, inducing three basic Frank elastic deformations in the director structure, namely, splay, twist and bend. The contribution of each deformation to the overall energy density F is given by [19, 125]

$$F = \frac{1}{2}\left[K_{11}\left(\nabla.\vec{n}\right)^2 + K_{22}\left(\vec{n}.\nabla \times \vec{n}\right)^2 + K_{33}\left(\vec{n} \times \nabla \times \vec{n}\right)^2\right] \tag{23}$$

where the proportionality constants K_{11}, K_{22} and K_{33} are associated with splay, twist and bend deformations, respectively. It is usually not possible to pack the nematic director field into curvatures without creating one or more defects. These defects are responsible for the transformation of LC droplet from one configuration to another. Different defect structures are classified on the basis of their two-dimensional structure often known as "strength (s)" of the defect. "s" is defined by the rotation of the nematic director on a closed path around the defect; s indicates how many times of 2π the director rotates. Since $+n$ and $-n$ of the director are equivalent, half-integer values of s are allowed. The light scattering and dielectric properties of different droplet configurations can vary considerably, which is an important factor in the PDLC devices. Accordingly, the configuration of confined LCs is an area of both scientific and technological interest [19, 126]. The configuration adopted by the nematic director field within a droplet reflects the subtle interplay of a number of forces, such as shape and size of cavity containing LC material, alignment properties of the LC at the polymer surface, elastic constants of the bulk nematic, temperature and the presence of external fields. Out of these factors, preferred alignment of the nematic at the surface of the polymer-LC interface determines the droplet configuration. If the anchoring energy is stronger than the elastic forces inside LC droplets, then the nematic director adopts a uniform tilt angle either (0° or 90°) at all points on the droplet surface, and the final configuration of nematic director within LC droplet is to minimize the total free energy. However, if the anchoring energy is weaker than the elastic forces inside LC droplets, then the tilt angle of the nematic varies spatially within the droplet to minimize curvature in the bulk of the droplet. Strong or weak anchoring conditions depend on the chemical nature of the polymer interface up to a certain degree. However, typically anchoring energy is more influential. Anchoring effects are magnified in small droplets, because of their shorter length scale and increased surface-to-volume ratio. Along with anchoring energy, balance of elastic constants is an imperative factor in determining the director configuration. Contribution of elastic constants to the system's free energy determines whether director configuration inside the droplet is simple or complex. The shape and size of cavity affects droplet structure. In submicron-sized droplets, the close proximity of surfaces and defects can distort the nematic structure throughout the droplet, whereas in large-sized droplets, it is easy to form multiple defect structures. The director configuration is isomorphic in a symmetric cavity and unpredictable in irregular-shaped cavity. The presence of external fields may influence the alignment direction of a nematic without altering the director configuration. Four commonly found director configurations in PDLCs [127–129] are illustrated in **Figure 32**.

a. Radial configuration (**Figure 32(a)**): In this configuration, director field is anchored perpendicular (homeotropic) to the droplet wall. A radial droplet possesses spherical symmetry, and the only elastic deformation present in this

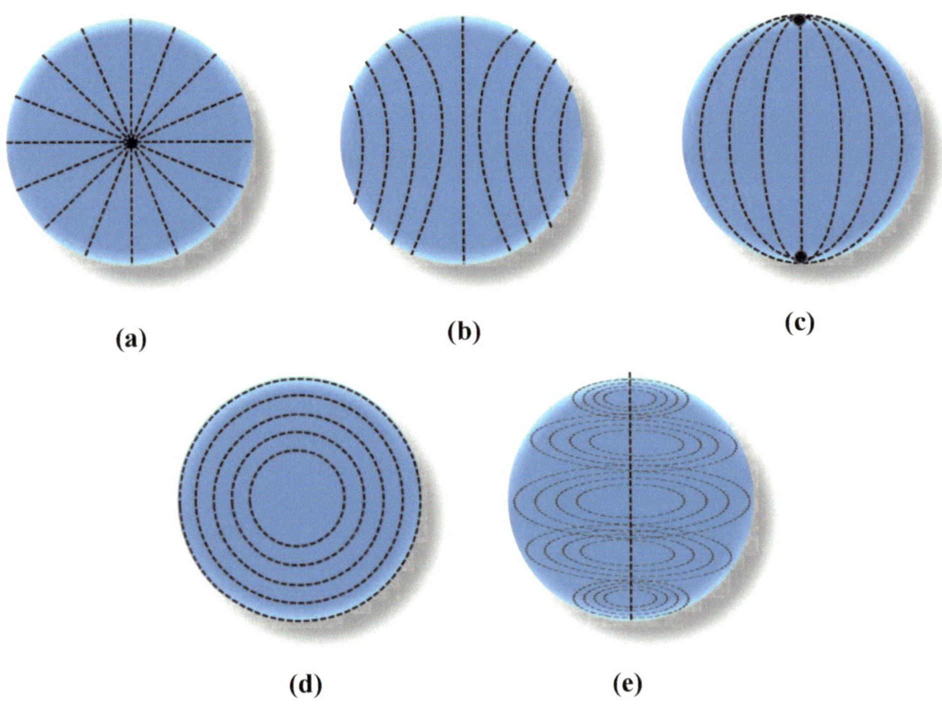

Figure 32.
Director configurations in a droplet of PDLC (a) radial, (b) axial, (c) bipolar, (d) toroidal and (e) three-dimensional view of the toroidal configuration.

structure is a splay deformation with a point defect (s = 1) in the volume centre known as hedgehog [19, 130]. This is shown in **Figure 33**. The defects (also known as disclinations) arise when the elastic energy density of a nematic grows sufficiently large, and the orientation of the nematic director becomes indistinct [131]. Radial structure is generally found by dispersing LC in low-surface energy fluids like polysiloxane or in glycerine containing a small amount of lecithin.

b. Axial configuration (**Figure 32(b)**): This configuration was found in director field which are weakly anchored perpendicular (homeotropic) to the droplet wall, smaller in radii than radial droplets and in those radial droplets which were exposed to electric or magnetic field [132]. The director field possesses cylindrical symmetry, with a line defect perpendicular to the preferred orientation direction at the droplet equator as shown in **Figure 32(b)**.

c. Bipolar configuration (**Figure 32(c)**): In this configuration, director field is anchored tangential (parallel or homogeneous) to the droplet wall. The director field possesses cylindrical symmetry, with the symmetry axis defined by two-point defects called *boojums*, which lie at opposite ends (poles) of the droplet. *Boojums* can exist only at the surface and cannot move into the volume of the droplet. In a bipolar droplet, both splay and bend deformations are present, with splay-type *boojums* located at the surface near the poles of the droplet, as shown in **Figure 33**. The bend deformation dominates throughout the rest of the drop along the lines connecting the two poles.

d. Toroidal configuration (**Figure 32(d)** and **(e)**): In the case of multiple elastic constants, the toroidal configuration exists when the splay energy becomes too large in comparison with the bend energy in the droplet. For a value of $K_{11}/K_{33} > 0.7$, a droplet with the toroidal configuration is a stable structure as it has a lower free energy than a bipolar structure. The toroidal structure possesses a line defect running along the droplet diameter, and the nematic director is everywhere perpendicular to this line arranged in a series of concentric circles.

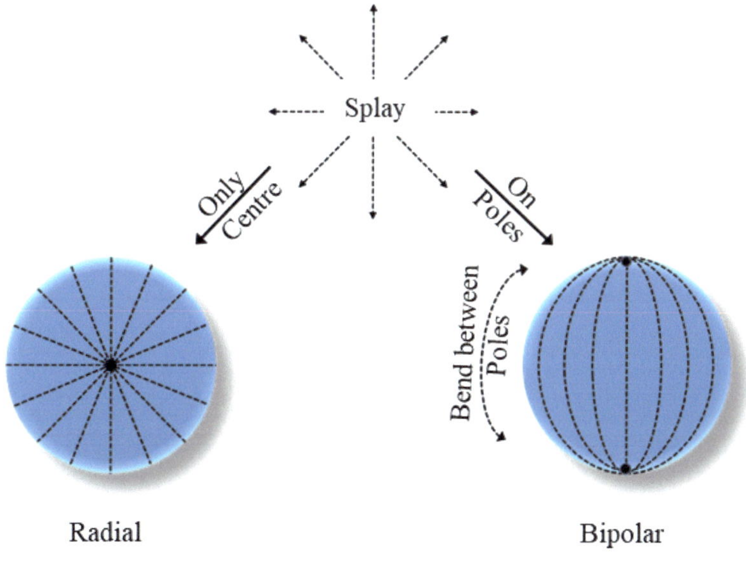

Figure 33.
A schematic illustration of splay and bend elastic deformations dictating the nematic ordering in radial and bipolar LC droplets.

6.3 Light scattering properties of PDLC films

Light scattering properties of PDLC films which can be controlled by electric field is a topic of great interest for scientific and technological reasons. The light scattering effects are insensitive to the initial polarization of light, and hence PDLC films can modulate light without the use of alignment layers or polarizers. The light scattering property of a PDLC film depends on many parameters such as LC droplet shape and size, droplet configuration, droplet density, refractive indices of LC and polymer, wavelength, etc. However, the nematic director orientation within the LC droplet dominates the light scattering properties of the films. Upon application of electric field, directors can be oriented along the direction of electric field causing transformation from highly scattering to highly transparent film. For simplicity, to explain the scattering and/or propagation of light at different voltage levels, we have developed a single droplet model [116, 133, 134].

In the absence of electric field, the different droplets will have different orientations. The droplet under investigation (**Figure 34(a)**) will scatter the light at polymer-LC interface. When a little voltage, below threshold voltage V_{TH} (defined later) (**Figure 34(b)**) is applied, because of the strong anchoring at interface, the alignment of LC droplet directors will not change much except the internal portion of the droplet. The internal portion experiences the electric field effect and aligns LC directors along the direction of the field. When the field (**Figure 34(c)**) is further increased up to the intermediate level, the majority of the LC droplets gets oriented along the direction of applied electric field, except the LC directors, which are on the polymer-LC interface, experiencing enough anchoring forces. At adequately high electric field (**Figure 34(d)**), i.e. above saturation voltage V_{ON} (defined later), all the directors get aligned along the direction of electric field. In such situation, light encounters only ordinary RI of LC, which is very close with the RI of the polymer. Therefore, a clear transparent film is observed at sufficiently high voltages.

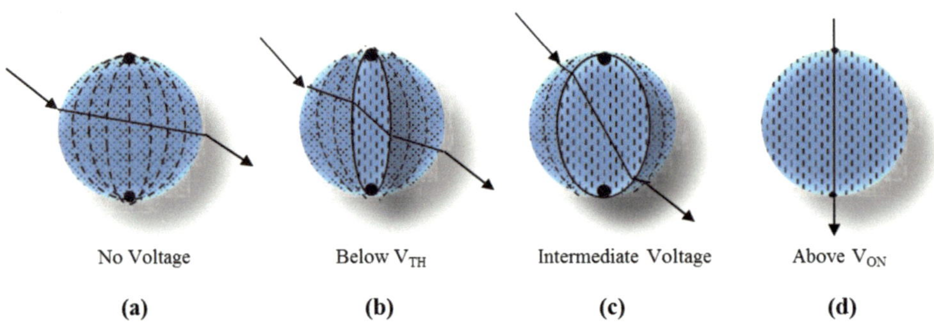

Figure 34.
Single LC droplet model: (a) V=0, (b) $V<V_{TH}$, (c) $V_{TH} \leq V \leq V_{ON}$, (d) $V>V_{ON}$.

6.4 Electro-optic properties of PDLC films

6.4.1 OFF-state transmittance (scattering) (T_{OFF})

In the absence of field, i.e. OFF state, the directors of LC droplet are randomly aligned. In this situation, most of the light incident normal to the film surface gets scattered giving opaque appearance to the film. The reason behind this is the mismatch of RI (a) between two adjacent droplets, (b) between a LC droplet and polymer and (c) within a LC droplet [135]. The OFF-state transmittance of a PDLC film depends on the birefringence of the LC. To minimize the value of OFF-state transmittance, the difference between LC extraordinary and ordinary refractive indices $\Delta n = n_e - n_o$ must be as large as possible.

6.4.2 ON-state transmittance (T_{ON})

In the ON state, when sufficient voltage is applied across the PDLC which overcome the anchoring at polymer-LC interface, the LC droplets attain minimum free energy by completely aligning themselves parallel to the field direction by the action of dielectric torque. In such a situation, light incident normal to the film surface experiences RI, $n_p = n_o$, and gets transmitted through the film.

6.4.3 Contrast ratio (CR)

Contrast ratio is the term used to evaluate display properties of PDLC films. It is the ratio of the ON- to OFF-state transmittance:

$$CR = \frac{T_{\text{ON}}(\%)}{T_{\text{OFF}}(\%)} \tag{24}$$

6.4.4 Transmittance difference (ΔT)

Another term used to evaluate the efficiency of PDLC films is transmittance difference (ΔT); it is difference of ON- and OFF-state transmittance:

$$\Delta T(\%) = T_{\text{ON}} - T_{\text{OFF}} \tag{25}$$

6.4.5 Switching voltages

One of the most important parameters of PDLC films is the voltage required to achieve an electro-optic effect. Initially the LC droplets are randomly oriented in PDLC films. When low electric field is applied, LC droplets starts orienting along the field; the voltage required to increase the transmittance of PDLC film by additional 10% of OFF-state transmittance (T_{OFF}) is termed as threshold voltage (V_{TH}). The theoretical model for threshold voltage has been developed by balancing the elastic forces, surface interaction and applied electric force and is mathematically derived as [136, 137]:

$$V_{\text{TH}} = \frac{d}{3a}\left[\frac{\rho_p}{\rho_{LC}} + 2\right]\left[\frac{k(l^2 - 1)}{\Delta\varepsilon}\right]^{\frac{1}{2}} \tag{26}$$

where d is the film thickness; ρ_p and ρ_{LC} are the resistivities of polymer and LC, respectively; k is the elastic constant; the aspect ratio $l = a/b$, where a and b are the length of the major and minor axes of LC droplet, respectively; and $\Delta\varepsilon$ is the dielectric anisotropy of the LC. With further increase in voltage, more and more LC align along the field. When sufficient voltage is applied across the PDLC which overcome the anchoring at polymer-LC interface, the LC droplet attains minimum free energy by completely aligning itself parallel to the field direction. At this stage film becomes fully transparent, and corresponding voltage is termed as saturation voltage or ON-state voltage (V_{ON}).

6.4.6 Response time

Another decisive factor in evaluating the performance of the polymer-LC composite film is its dynamic response to an applied electric field. Quick response of a

PDLC film is critical in many applications. Response time is the time required by the LC molecules to align along the electric field upon application of field and to relax to their initial orientation when the electric field is removed. It generally depends on the relative strength of the applied field and the elastic reorientation forces. It can be affected by the degree of phase separation between nematic and polymer phase. Polymer-LC material properties and film morphology which include droplet size, shape, multiple scattering processes, etc. have high impact on response time of composite film [138, 139]. Erdmann observed reduced values of response time for elongated LC droplets [128]. It is calculated by balancing an elastic torque (Γ_d), electric torque (Γ_e) and viscous torque (Γ_v). It is the sum of rise time (τ_r) and decay time (τ_d). Rise and decay times are usually defined in terms of an optical response. Rise time is the time in which transmittance of film reaches from 10 to 90% on application of electric field. Similarly decay time is the time in which transmittance reaches from 90 to 10% on removal of electric field [115]. Mathematically

$$\frac{1}{\tau_r} = \frac{1}{\gamma_1}\left[\Delta\varepsilon \times V^2 + \frac{k(l^2 - 1)}{a^2}\right] \tag{27}$$

and

$$\tau_d = \frac{\gamma_1 \times a^2}{k(l^2 - 1)} \tag{28}$$

where $\gamma_1$is the rotational viscosity of LC and other symbols manifest same meaning defined earlier.

For higher electric fields

$$\tau_r = \frac{\gamma_1}{\Delta\varepsilon \times V^2} \tag{29}$$

An analysis of the above equation indicates that τ_r is predominantly a function of the applied voltage, where τ_d typically depends on the configuration of LC domain and their anchoring energy with polymer wall. A good estimation of rise and decay times can be attained by understanding the director direction inside LC droplet.

6.4.7 Hysteresis effect

During the study of electro-optic properties, a well-known hysteresis phenomenon was observed in PDLC films [138]. The hysteresis is a known problem, which must be addressed for practical application of the display material. It has been found that the transmittance values obtained at various voltages during scan down cycle do not follow the same path as that of the scan up cycle. A measure of hysteresis is given by the voltage width at half of maximum transmittance (ΔV_{50}). The lag of transmittance during scan down cycle may trace on any of the three paths as shown in **Figure 35**:

1. Hysteresis: Hysteresis occurs in intermediate voltage range, and the transmission at a given voltage depends on the previous voltage state. The situation when a higher transmission at a given voltage is observed during the scan down cycle as compared to the transmission at the same voltage during the scan up cycle is termed as hysteresis effect.

2. Persistence: The phenomenon when the film does not return to its complete scattering state immediate after removal of field is termed as persistence. The reason behind persistence is high interconnectivity between LC droplets which creates defect structure in any of the droplet. Then the other connected droplets may get trapped in the high field and remain in the same state even after the removal of field, until the defect escapes from the trap.

3. Memory effect: It is a semi-permanent persistence because film remains permanently in the transparent state even after removal of field. The reason for memory effect is weak anchoring force which develops between the boundaries of polymer network and LC domain upon application of field.

There are some common factors in hysteresis, persistence and memory effect in PDLC films. Persistence and memory effects are due to predominance of some specific factor.

The sources which give rise to hysteresis phenomenon have been studied by various researchers [117, 135, 138, 140, 141]. Hysteresis depends on the rate at which fields are applied and removed. For quickly (period of microseconds) varying voltages, hysteresis are more as compared to slowly varying (period of seconds) voltages in a same film [19]. It has been proposed that the orientation mechanism of the LC droplet director is a crucial factor responsible for hysteresis. Upon application and removal of field, there is substantial difference in the director direction of LC molecules, which are at the polymer-LC interface and which are inside the LC droplet [138]. The orientation-reorientation of LC directors depends on the polymer-LC compatibility and induced interfacial polarization, influencing the distribution of the relaxation time [136]. Hysteresis might be due to the presence of defect structure within a LC droplet [138, 141]. Depolarization effects might contribute to hysteresis through the generation of electric fields inside the PDLC films [142]. It has been suggested that hysteresis might be due to the residual electric charge (on field removal) which serves as a capacitor during the scan down cycle [84, 135]. On analysing **Figure 35**, it is clear that hysteresis effect is not observed at high fields because at higher voltages of scan down cycle, LC droplets

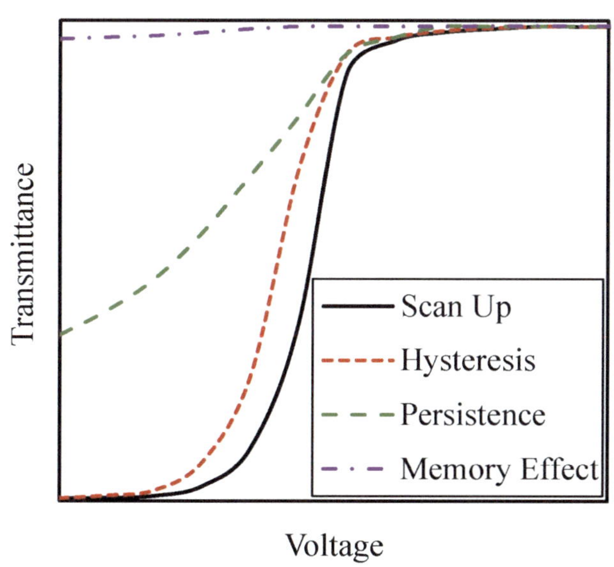

Figure 35.
Schematic of hysteresis, persistence and memory effect in a PDLC film.

remain in the same state of orientation. However, when the applied field is reduced further, reorientation of LC droplets director begins, giving rise to hysteresis effect [143].

There is a much scope of improvement in the optical and dielectric properties of PDLC devices, by reducing operating voltages, quickening switching and relaxation times, enhancing image contrast, etc. Before photopolymerization, PDLC films are a homogeneous mixture composed of monomer and LC which transforms into heterogonous system consisting LC droplets embedded in polymer matrix after photopolymerization.

Several groups have suggested that to augment the PDLC film properties, we can alter:

1. Film preparation conditions, such as, reaction temperature, UV irradiation intensity, curing time and film thickness

2. Properties of monomer unit such as their structure, functionality, concentration in LC, etc.

3. Physical properties of LC such as optical and dielectric parameters, rotational viscosity, temporal characteristics, molecular dynamics, etc. [76, 119, 120, 123, 144–149].

To enhance the properties of PDLC films, LC properties are modified suitably by addition of nanoparticles, dyes, polymers and carbon nanotubes. The inclusion of new additives can add new functionalities to the obtained devices. To obtain the desired result from the host LC material, proper selection of size, shape and structure of guest/additive particle is an imperative factor. In has been proved that due to the inherent dipole moment, LC materials respond radically when doped and anchored with elongated species.

The incorporation of small quantity of dye molecules to the LC is one suitable solution, which has been studied by several researchers.

6.5 Operating principle of DPDLC composite films

DPDLC films are also named as guest-host polymer-dispersed liquid crystal (GHPDLC) composite films. The LC phase acts as the host and dye molecules doped in LC act as the guest. Once the geometrically anisotropic dye molecules are added/dissolved in LC, the long molecular axis of dye molecule tends to align along the LC director; this is called as guest-host interaction. Dyes used for LC media must have high dichroic ratio, high-order parameter, high stability and good compatibility and solubility with LC but not with monomer unit or polymer matrix. It must be a positive type, i.e. the absorption transition dipole is along the long molecular axis. When the polarization of the incident light is perpendicular to the long axis of the dye molecules, the light is weakly absorbed. When the polarization of the incident light is parallel to the long axis of the dye molecules, the light is strongly absorbed, as shown in **Figure 36**. Considering all the above-mentioned points, dichroic azo dyes, which are N=N substance and absorb light of certain wavelength more along one direction than the other, have been used.

Similar to the PDLC films, the dispersion of the droplets can also be achieved using polymerization-induced phase separation method, in which the dye and LC mixture separate from the polymer binder. These DPDLC films possess controllable

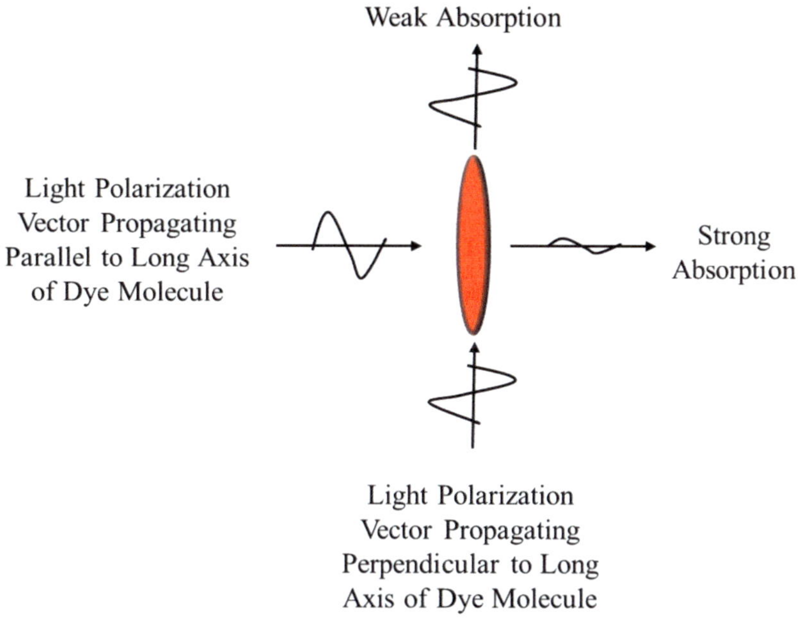

Figure 36.
A schematic representation of direction dependent absorption of dichroic dye molecule.

scattering as well as controllable absorbance, modulated by the liquid crystal molecule. The large transition moment of the rod-shaped dye molecules is modulated by the symmetry axis of the LC molecule in the OFF and ON states of electric field [150–153]. In the field "OFF state", dye molecules are randomly oriented along with the LC droplets. The unpolarised incident light is scattered as well as absorbed due to LC droplets and dissolved dye molecules. Due to the scattering of light, the optical path length of light increases, and hence enhancement in absorption has been noticed. In the field "ON state," LC droplets and dissolved dye molecules get aligned along the field direction, making the film feebly absorbing; hence, most of the light is transmitted through the film. **Figure 37** shows the schematic of DPDLC film in the OFF and ON states of applied electric field [61, 125].

LC droplets in all the composite films exhibit predominantly bipolar (LC droplet with cylindrical symmetry axis defined by two-point defects at opposite ends)-type morphology with some radial (LC droplet with spherical symmetry and the only point defect in the volume centre) structures. No preferred orientation of LC droplets was found, which is responsible for the scattering of light. Upon addition of dichroic dye molecule, significant morphological changes arise in LC droplet. The dye, polymer and LC are chosen such that the solubility of dye is high in LC but very low in polymer. **Figure 38** shows POM image of DPDLC film (O36N), prepared using LC BL036 and prepolymer NOA65 in wt/wt% with varying disperse Orange 25 dye concentration. As the dye concentration increases from 0–1%, there is noticeable increase in droplet size from submicron to ~10 μ. The reason is dye molecules dissolved in LC absorb some of the UV radiations provided for photopolymerization, which reduces rate of polymerization and in turn influences droplet growth. At lower dye content, faster polymerization rate produces smaller and denser LC droplets. At higher dye content, slow polymerization rate preserves polymer in a liquid state for an extended period, thus allowing the growth and coalescence of small LC droplets to form bigger ones. There is more number of bipolar droplets in the high dye-doped film

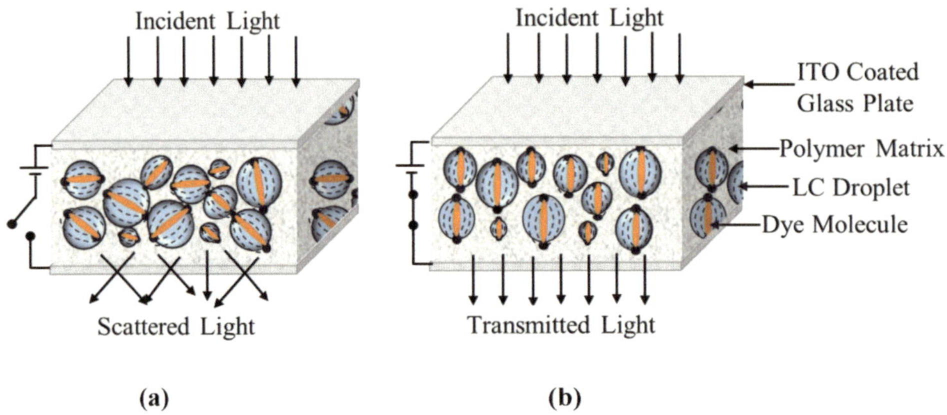

Figure 37.
Schematic of DPDLC in (a) OFF state and (b) ON state.

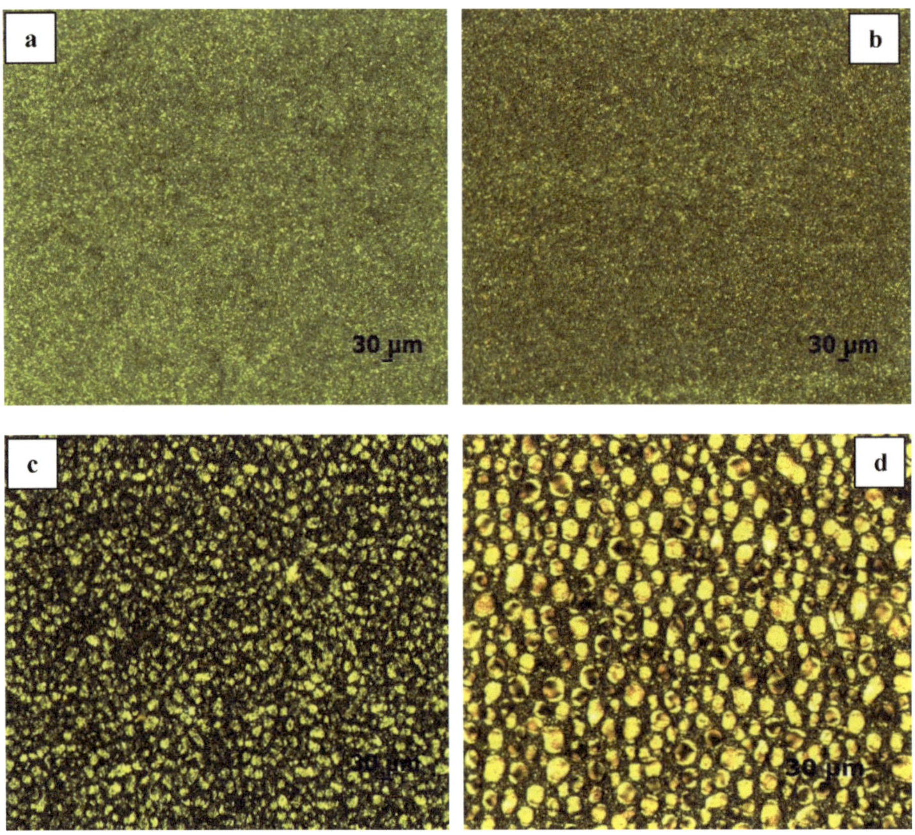

Figure 38.
POM images of O36N DPDLC film with dye concentrations (a) 0%, (b) 0.015%, (c) 0.25% and (d) 1%.

(**Figure 38(d)**) showing relatively higher affinity for the polymer and thus aiding strong anchoring at the polymer-LC interface than the low dye-doped composite films. Altogether, the addition of dye increases the viscosity of polymer-LC mixture, especially at low temperatures, which increases droplet size and hence optical path length [143, 154–158].

The high dye-doped PDLC film emerges with more number of bipolar droplets, as shown in **Figure 38(d)**. This can be interpreted as high dye-doped LC molecules show relatively higher affinity for the polymer than the low dye-doped LC molecules and hence facilitating strong anchoring conditions at the interface of polymer and LC molecules [155].

Figure 39 shows effect of electric field on the morphology of DPDLC film O36S prepared using disperse Orange 25 dye, LC BL036 and Prepolymer SAM-114. The LC and prepolymer are taken here in 45/55 wt/wt%. In the ON state, when the electric field is low (10–20 V) (**Figure 39(b)**), there is substantial difference in the director direction of molecules which are near the polymer-LC interface and which are inside the LC droplet, and point defects are still at their initial position. On increasing the voltage (higher than anchoring energy at polymer-LC interface) (30 V and above) (**Figure 39(c)** and **(d)**), small and big LC droplets acquire maltese-type structure and twisted arrangement with director direction parallel to field, respectively. The appearance of maltese crosses is due to the interference (recombination) of the two refracted waves at higher fields [125, 159, 160].

The voltage dependence of transmittance for DPDLC composite film (O00N), prepared using dye disperse Orange 25+ LC HPC850100–100 + prepolymer NOA-65, at a constant frequency of 200 Hz is shown in **Figure 40**. LC and prepolymer are taken in 40/60 wt/wt% ratio, respectively, with dye content varying from 0 to 1%. The parameters such as T_{OFF}, T_{ON}, CR, ΔT, V_{TH} and V_{ON} calculated from graph (**Figure 40**) are summarized in **Table 4**.

From the above graph (**Figure 40**) and **Table 4**, it is clear that the DPDLC film with lowest dye concentration (0.007%) is the optimum one. This film has low value of T_{OFF} and high value of T_{ON}, which results in high CR and high ΔT. The value of V_{TH} and V_{ON} is remarkably low, desired for good EO devices.

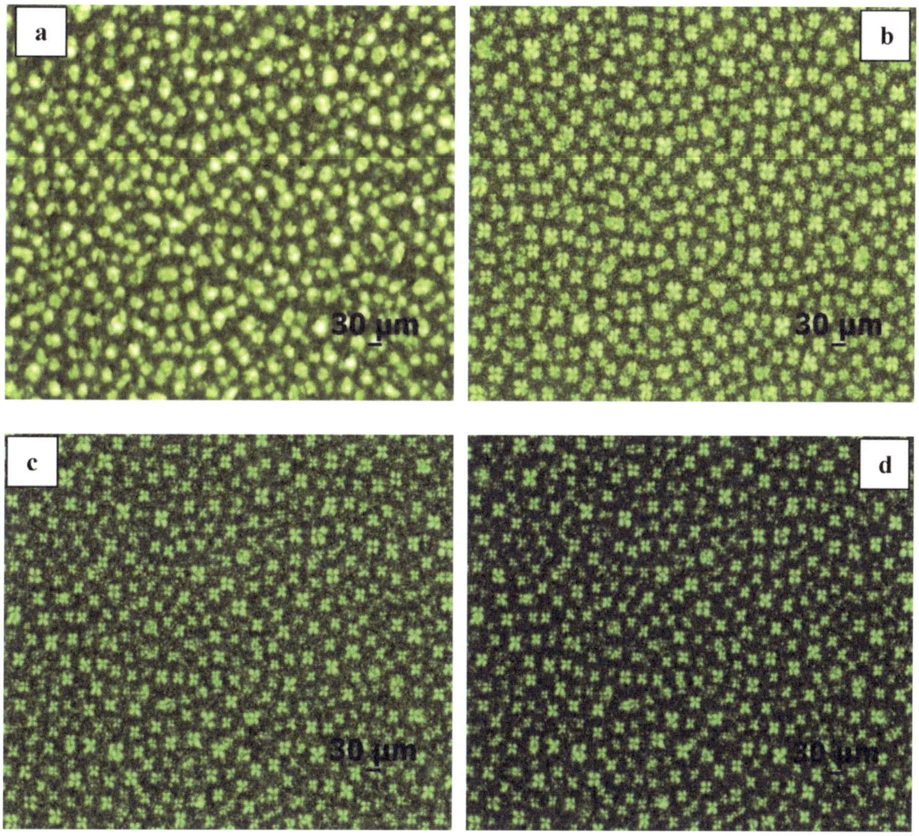

Figure 39.
Droplet structure of 0.015% O36S DPDLC film at (a) 0 V, (b) 20 V, (c) 40 V and (d) 60 V.

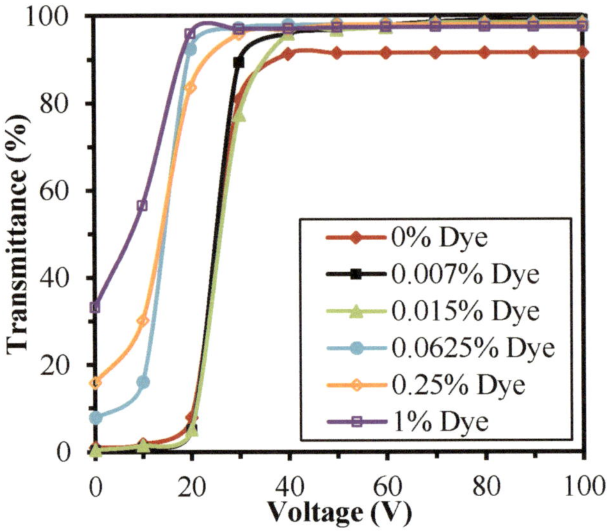

Figure 40.
Transmittance vs. voltage curve of various dye concentrations OooN DPDLC films at 200 Hz frequency and 25°C temperature.

Dye concentration (wt%)	T_{OFF} (%)	T_{ON} (%)	ΔT (%)	CR	V_{TH}	V_{ON}
0.0	1.02	91.43	90.41	90	3.5	40
0.007	**0.41**	**98.98**	**98.57**	**241**	**1.1**	**40**
0.015	0.51	98.37	97.86	193	1	40
0.0625	7.86	97.96	90.1	12	1.5	29
0.25	15.52	97.96	82.44	6	1.3	33
1	33.16	97.40	64.24	3	1.5	29

The significance of bold values are the optimum parameters for the given film.

Table 4.
Voltage-transmittance data of various dye concentrations OooN DPDLC films.

The variation in rise time and decay time of DPDLC composite films (O36S) prepared using LC BL036+ Prepolymer SAM-114 in 45/55 wt/wt% ratio and doped with disperse orange dye, at 40 V, is given in **Table 5**.

Generally, the rise time of all composite films decreases with the increase in voltage, whereas decay time is independent of voltage. This observed phenomenon is in good agreement with Eqs. (27) and (28). However, sometimes, there is small increase in decay time with voltage indicating longer transparent state even after field removal. Also, at higher voltages, rise time is of the order of microseconds indicating quick switching of the films. **Table 5** indicates lower value of rise time for low dye concentration DPDLC film. The proposed reason is increased dipole

Dye concentration (wt%)	Rise time (ms)	Decay time (ms)
0	0.4	6.8
0.007	0.2	6.6
0.015	0.3	6.7
1	0.8	8.3

Table 5.
Rise time and decay time of DPDLC film O36S at 40 V.

moment of LC droplet due to addition of dye molecules, contributing in LC droplet quick orientation, hence low rise time. With the further increase in dye concentration, rise time increases because of increase in viscosity of LC+ dye mixture. Altogether, at higher dye concentration, LC molecules expel dye towards polymer surface creating additional surface anchoring. Upon removal of field, LC droplets of low dye concentration DPDLC film quickly reorients as compared to the un-doped film. In high dye concentration DPDLC film, LC droplets took more time to reorient because of the additional surface anchoring between dye-doped LC droplet and polymer walls. During application of field, charge gets stored in dye-doped LC droplet, which serves as capacitor even after removal of external field affecting relaxation time of LC droplet.

For the analysis of experimental data and qualitative evaluation of distribution of relaxation time, Cole-Cole plot is drawn for imaginary part ε″ vs. real part ε′ of the dielectric constant. For O0ON (dye Orange 25+ LC HPC850100–100 + prepolymer NOA65) DPDLC film, Cole-Cole plot is shown in **Figure 41**. Dielectric parameters calculated using Cole-Cole plot are summarized in **Table 6**. The zero value of α listed in **Table 6** clearly shows that DPDLC films show Debye type behaviour. The 0.007% DPDLC film has very high value of dielectric strength.

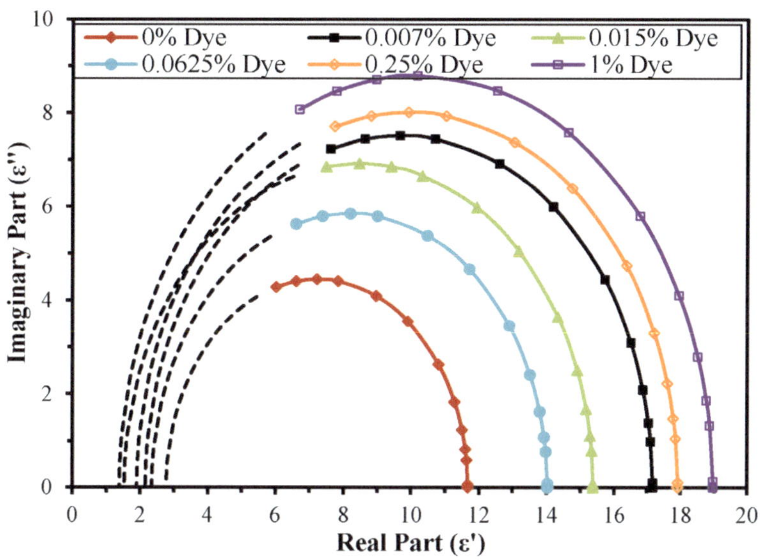

Figure 41.
Cole-Cole plots of various dye concentrations OooN DPDLC films at 25°C temperature.

Dye concentration (wt%)	f (MHz)	ε_s	ε_∞	$\delta\varepsilon'$	τ (s)	α
0.0	15.1	11.66	2.77	8.89	1.05 E^{-8}	0
0.007	**15.1**	**17.17**	**2.15**	**15.02**	**1.05 E^{-8}**	**0**
0.015	17.4	15.36	1.53	13.83	9.15 E^{-9}	0
0.0625	15.1	14.03	2.34	11.69	1.05 E^{-8}	0
0.25	15.1	17.91	1.89	16.02	1.05 E^{-8}	0
1	13.2	18.96	1.38	17.58	1.21 E^{-8}	0

Table 6.
Fitting parameters of various dye concentrations OooN DPDLC films.

In order to improve the performance of PDLC films, similar to DPDLC films, LC can be doped with nano-particles or carbon nanotubes. One such example is multiwalled carbon nanotubes (MWCNT)-doped PDLC (CPDLC) film, which is discussed here.

6.6 Operating principle of CPDLC composite films

Similar to the DPDLC composite films, in CPDLC films, LC phase acts as the host and MWCNT doped in LC act as the guest entity. MWCNTs are physically and environmentally stable, mechanically strong, chemically inert and thermally and electrically conducting material with high aspect ratio. MWCNT is also self-organizing, like LC material, but strong attractive Van der Waals forces between adjacent MWCNT lead them to cluster and form unorganized bundles [161]. It is very difficult to disperse MWCNT in a medium, but small percentage of MWCNT can be dispersed well in LC fluid. Till date any experimental confirmations regarding the interaction between MWCNT and LC are unavailable, but on analyzing their structure, π-π interaction (aromatic interaction) between MWCNT walls and phenyl rings of LC molecules is evident. As MWCNT is insoluble in LC, this type of interaction is quite weak to cause any LC director deformation. The well-dispersed MWCNT are generally orientated with their cylindrical axis parallel to the director direction of nematic LC. After the sufficiently stable dispersion of MWCNT in LC, both host LC and guest MWCNT share their intrinsic properties with each other, which are listed below:

1. Due to the dipolar nature of LC material, asymmetric charges are induced on MWCNT producing permanent dipole moment on MWCNT.

2. Nematic LC solvent provides partial orientational order to MWCNT.

3. MWCNT impart their electrical conductivity to the LC molecules.

Because of this sharing of inherent properties with each other, properties of CPDLC films get affected significantly. Similar to the DPDLC films, in CPDLC films also the dispersion of the LC droplets has been achieved using PIPS method, but in CPDLC films even after completion of polymerization process, MWCNT are found to be well separated entity from LC droplets embedded in polymer matrix. Since MWCNT do not absorb any light, hence operating principle of CPDLC films is based only on controllable scattering of light from randomly dispersed LC droplets. In the absence of field, i.e. "OFF state", MWCNT and LC droplets are separately and randomly oriented inside polymer matrix as shown in **Figure 42(a)**. The unpolarised incident light is scattered because of LC droplets. Upon application of field, i.e. "ON state", LC droplets get aligned along the direction of field, and MWCNT also get partially oriented as shown in **Figure 42(b)** and may form conducting channel at higher MWCNT concentration.

Figure 43(a)–(d) shows SEM images of some representative CPDLC films (C00N) prepared using LC HPC850100-100 and prepolymer NOA-65 doped with MWCNT concentration 0, 0.005, 0.05 and 0.5%. Here, LC and prepolymer are taken in 60/40 wt/wt% ratio. As MWCNT are insoluble in LC and do not participate in photopolymerization kinetics, therefore, the size of the LC droplets remains invariant upon addition of MWCNT. On analysing SEM images of 0% and 0.005% CPDLC films (**Figure 43(a)** and **(b)**), it is clear that the size of cavities is of few microns. Increase in MWCNT concentration (0.05 and 0.5% CPDLC film) does not affect the size of LC droplets but because of clustering of MWCNT, few bigger size cavities are formed as shown in **Figure 43(c)** and **(d)**.

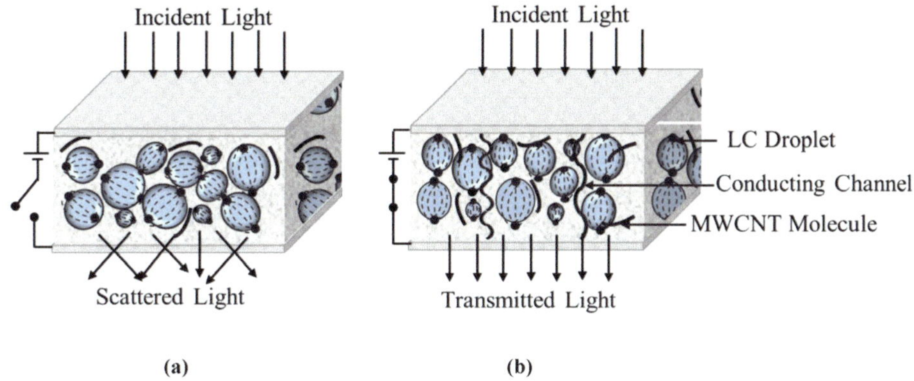

Figure 42.
Schematic of CPDLC in (a) OFF state and (b) ON state.

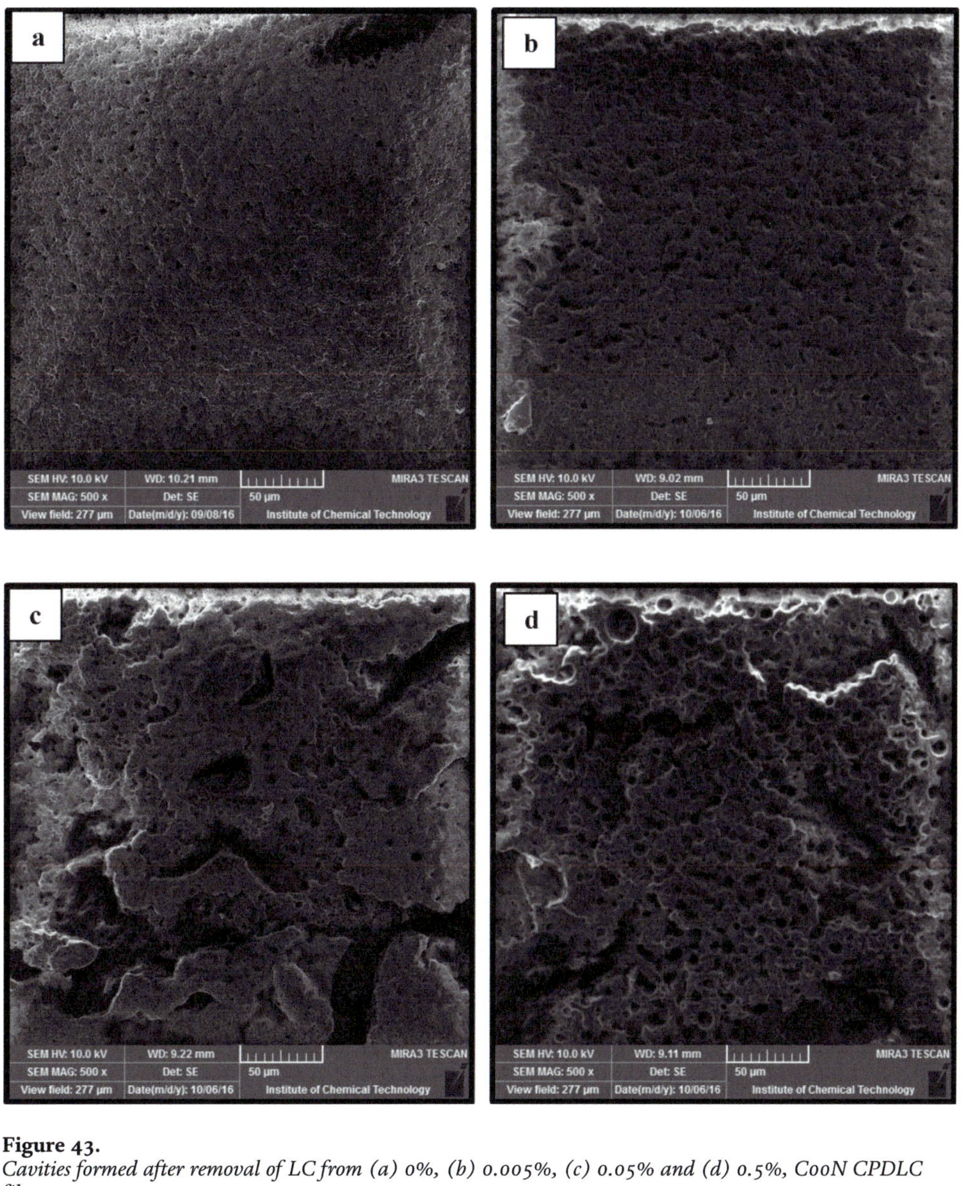

Figure 43.
Cavities formed after removal of LC from (a) 0%, (b) 0.005%, (c) 0.05% and (d) 0.5%, CooN CPDLC films.

The voltage dependence of output transmittance for CPDLC films with MWCNT content varying from 0 to 0.5% is shown in **Figure 44**. The parameters such as T_{OFF},T_{ON}, CR, ΔT, V_{TH} and V_{ON} calculated from this graph are summarized in **Table** 7.

From the above graph (**Figure 44**) and **Table** 7, it is clear that the CPDLC film with lowest MWCNT concentration (0.005%) is optimum one. This film has low value of T_{OFF} and high value of T_{ON}, which results in high *CR* and high ΔT. The value of V_{TH} and V_{ON} is remarkably low, desired for good EO devices.

In order to understand the effect of temperature on other dielectric parameters such as dielectric strength and relaxation process, Cole-Cole plot can be drawn between real and imaginary parts of complex dielectric constant. **Figure 45** shows Cole-Cole plots of 0.005% MWCNT-doped C36N CPDLC film (LC BL036 and prepolymer NOA-65 in 50/50 wt/wt % ratio) at different temperatures. It is clear from graphs that the value of dielectric strength {difference of relative dielectric constant at static (at 20 Hz) and optical (at relaxation) frequency *f*} increases up to 40°C and then starts decreasing because small increase in temperature weakens intermolecular interaction and hence relaxes orientation-reorientation process of dipoles, whereas at high temperatures, thermal agitation becomes more predominant than intermolecular interaction which produces randomization of dipoles.

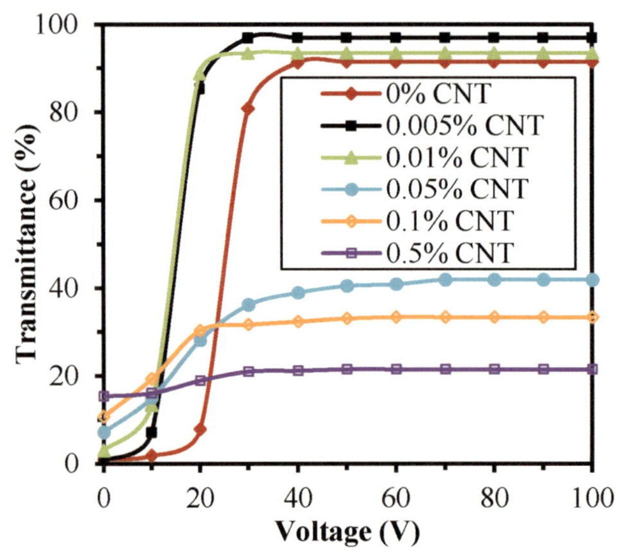

Figure 44.
Transmittance vs. voltage curve of various MWCNT concentrations C00N, CPDLC films at 200 Hz frequency and 25°C temperature.

MWCNT concentration (wt %)	T_{OFF} (%)	T_{ON} (%)	ΔT (%)	CR	V_{TH}	V_{ON}
0.0	1.02	91.42	90.4	90	3.5	40
0.005	**0.71**	**97.02**	**96.31**	**137**	**0.1**	**30**
0.01	3.06	93.46	90.4	31	0.4	24
0.05	7.14	41.83	34.69	6	1.0	32
0.1	10.71	33.36	22.65	3	1.3	20
0.5	15.3	21.42	6.12	1	7.4	30

Table 7.
Voltage-transmittance data of various MWCNT concentrations C00N CPDLC films.

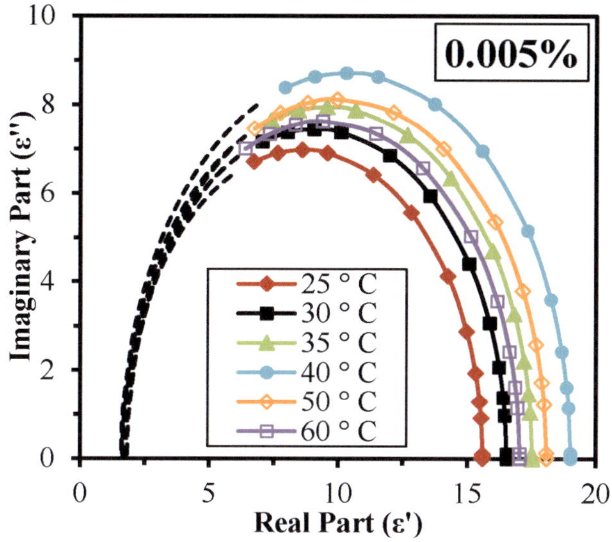

Figure 45.
Cole-Cole plots of 0.005% MWCNT concentration C36N, CPDLC film at different temperatures.

The value of distribution parameter α (not shown here) calculated for all Cole-Cole plots drawn for CPDLC films at different temperatures was zero, revealing Debye type relaxation [162].

6.7 Conclusions of PDLC study

The phenomenal optical and dielectric anisotropy of LC has been exploited in various display devices/LC technologies, and one such example is polymer-dispersed liquid crystal films. To improve the optical efficiency of PDLC device, with reduced operating voltages, faster switching time and high image contrast, properties of LC have been modified by doping it with some foreign entity. To obtain the desired result from the host LC material, proper selection of size, shape and structure of guest is an imperative factor. In all cases it has been proven that LC responds radically when doped and anchored to elongated species due to their inherent dipole moment. Therefore, host LC can be doped with dichroic dye guest molecules. Dye molecules tend to line up with the LC director, and dye absorbance is modulated by the alignment of nematic director with an external electric field. The controlled absorption and scattering of light through these materials make them promising candidates for various potential devices. Also the self-organizing properties of nematic LCs can be used to align carbon nanotubes (CNT) dispersed in them. CNT not only well integrate in the matrix but also, even at very low concentration, have a detectable effect on the LC properties that can be very attractive for display applications.

7. Holographic polymer-dispersed liquid crystal

When monomer concentration is high around 60–70%, nanosized LC droplets are formed and embedded inside the polymer matrix, this kind of polymer-LC composite films is named as HPDLC films. In PDLC films LC droplets are randomly distributed in polymer matrix, whereas in HPDLC films, alternate polymer-rich and LC-rich regions exist. As the size of the LC droplets is much smaller than the wavelength of visible light, composite films, free from scattering effect and with

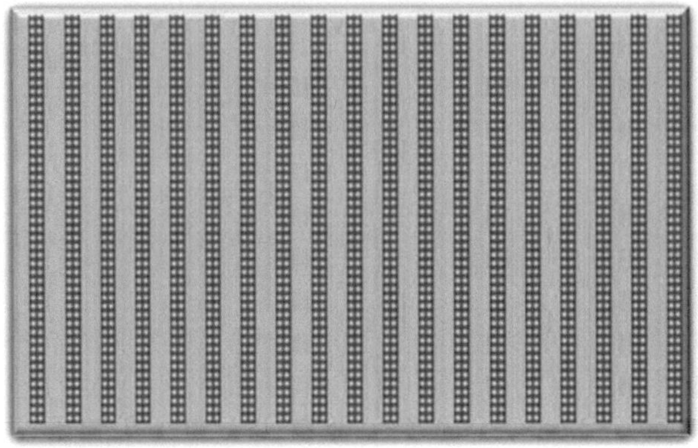

Figure 46.
Schematic of HPDLC film.

faster response time, have been obtained. Similar to the PDLC composite films, no surface alignment layer is needed in HPDLC films. **Figure 46** shows schematic of HPDLC film.

7.1 Operating principle

When a mixture of LC, monomer and photoinitiator (PhI) is exposed under the standing wave, formed from the interference of two or multiple coherent laser beams, it generates periodic dark and bright fringes. These periodic fringes regulate polymerization and hence phase separation process. High polymerization rate in the bright region (diffusion of monomer from dark to bright region) and low polymerization rate in the dark region (diffusion of LC from bright to dark region) create Bragg grating with alternate polymer-rich and LC-rich regions. Similar to the PDLC, the polymer and LC material are chosen such that the RI of the polymer should match with the ordinary RI of LC. Upon application of electric field, HPDLC film becomes optically transparent/homogenous, and the grating is in its OFF state. When the electric field is removed, LC molecules return to their original random state, and grating is in its ON state. This HPDLC grating reflects light of a particular wavelength and transmits light of all other wavelengths [163]. Morphology and diffraction properties of grating depends upon writing set-up, materials, diffusion rate, curing conditions and phase separation process [164, 165]. The particular wavelength that is reflected is a function of the refractive index difference and the width of the layers in the grating. When a voltage is applied, the liquid crystals align with the field, and their new refractive index matches that of the polymer, causing the grating to become transparent.

7.2 Material used and sample preparation of HPDLC films

HPDLC composite films are also prepared by mixing LC, monomer, photoinitiator and dopant (if any) in a desired ratio.

7.2.1 LC material

The optical and dielectric anisotropy of nematic LC make them suitable candidate for HPDLC films. Similar to the PDLC composite films, the operation of these composite films is based on birefringence property of LC. Nematic LCs are optically uniaxial materials, i.e. they have two direction-dependent refractive indices, (ordinary RI, n_o, and extraordinary RI, n_e) with birefringence $\Delta n = n_e - n_o$ and average RI, $n_{av} = \sqrt{\frac{1}{3}\left(n_e^2 + 2n_o^2\right)}$. The value of Δn may be positive or negative [34, 37, 38]. The polymer and LC are chosen such that the refractive index of the polymer (n_p) should match with ordinary refractive index (n_o) of the LC, typically $n_e \gg n_o \sim n_p$.

7.2.2 Monomer unit

In HPDLC films two types of monomer can be used:

a. Acrylate-based monomer: A wide range of acrylate-based monomers with different functionalities are available such as phthalic diglycol diacrylate (PDDA), 2-ethylhexyl acrylate (EHA), polyurethane acrylate oligomers (PUA), etc. But these multifunctional acrylates carry some drawbacks in HPDLC gratings, such as: (i) Because of early gelation and subsequent vitrification, monomer conversion is low, and polymerization can continue for prolonged periods, even in the dark; (ii) during exposure considerable shrinkage was also observed; (iii) larger interfacial area increases unwanted optical scattering; and (iv) interconnectivity in LC droplets and variety in LC droplet configuration decreases prompt electrical switching.

b. Thiol-ene-based monomer: Thiol-ene-based monomers are composed of multifunctional aliphatic thiols and vinyl monomers comprising ene groups. These monomers have high conversion efficiency, good stability, less shrinkage and more elastic; therefore they are preferably used for constructing HPDLC composite films. Norland optical adhesives are good examples of these kinds of monomers. Earlier these monomers could only be cured by UV light, but now new materials were synthesized, which can be cured under visible light.

7.2.3 Photoinitiator and co-initiators

Along with LC material and monomer, to induce photopolymerization, PhI and co-initiators are also required. Choice of PhI depends on the wavelength of laser beam employed for writing. Sometimes chain extenders are also incorporated into the mixture to optimize grating morphologies.

7.2.4 Sample preparation

Empty sample cell is prepared by two ITO-coated glass substrate (ITO coating facing each other) separated by suitable spacers. Mixture of LC, monomer, photoinitiator and dopant (if any) is stirred for homogenization. Mixture is filled into ITO cell and then exposed under suitable light depending upon the PhI. Generally, for curing sample cell is placed under the interference pattern formed by coherent laser beams. Samples are again placed under UV lamp for postcuring if needed [166].

7.3 Types of HPDLC gratings

Different writing set-ups produce different types of HPDLC gratings, named as transmission grating and reflection grating. If the writing beams are incident on

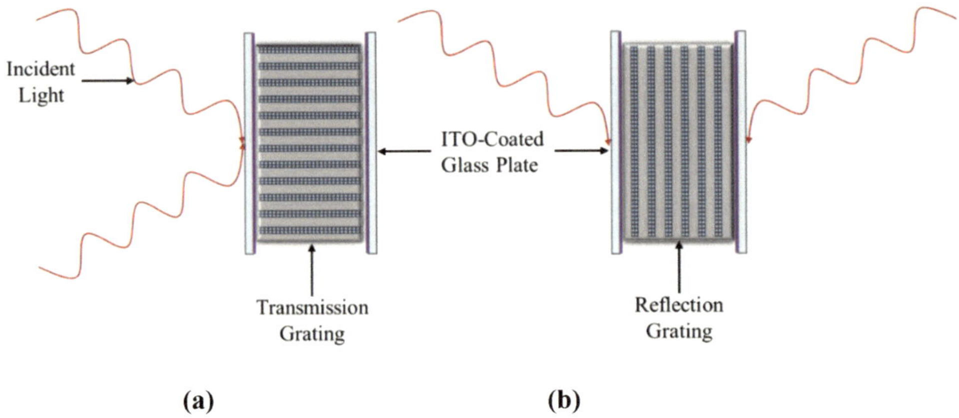

Figure 47.
Types of HPDLC grating: (a) transmission grating and (b) reflection grating.

the same side of sample cell, transmission grating will be formed, with grating planes perpendicular to the sample cell surface as shown in **Figure 47(a)**. If the writing beams are incident from both sides of the samples, reflection grating will be formed, with grating planes parallel to the sample cell surface as shown in **Figure 47(b)**.

7.4 HPDLC grating parameters

7.4.1 Grating period/grating pitch

The grating period depends on the writing wavelength and intersection beam angle:

$$\Lambda = \frac{\lambda}{2\overline{n}\ \sin\left(\theta_i/2\right)} \tag{30}$$

Here, Λ is the grating period, λ is the writing wavelength and $\overline{n}$ is the average refractive index of the material mixture. A grating with varied period can be obtained by inserting a refractive cylindrical lens in conventional double interference optical path. It varies the angle of incidence of one of the two interference laser beams [167].

7.4.2 Cook-Klein parameter

This parameter depends on the grating period and thickness of grating:

$$Q = 2\pi\frac{\lambda d}{\overline{n}\Lambda^2} \tag{31}$$

Here, λ is the wavelength of incident light, and *d* is the thickness of the grating.

If the value of $Q < 1$, then it is a Raman-Nath-type grating. It is thin grating and multiple diffraction orders can be found.

If the value of $Q > 1$, then it is Bragg-type grating. It is thick or volume grating, and only zero order or first order of diffraction can be found. Since the optical losses are low in Bragg-type-grating, it is preferred for practical applications.

7.5 HPDLC morphology

Different materials and varying curing conditions produce different types of polymer-LC morphologies in HPDLC composite films. Mainly, HPDLC morphologies can be categorized into three types.

7.5.1 LC droplet-like morphology

Generally, this type of morphology has been found in transmission or reflection grating (**Figure 48**). These types of gratings are made up of acrylate- or thiol-ene-based monomers. During PIPS process, because of fast curing process (intense curing light and small curing time) and high effective functionality of monomer, LC molecules get diffuse and configure themselves into distinct and elongated LC droplets. These LC droplets, embedded into the polymer matrix, are distinguished in SEM image (**Figure 48(b)**). Here, light scattering is more because the size of the LC droplets is comparable to the wavelength of the visible light [164, 168].

7.5.2 Polymer scaffolding morphology

In transmission gratings prepared from acrylate-based material systems under slow curing process, polymer scaffolding morphology can be obtained (**Figure 49**). Here instead of small LC droplets, large LC domains or LC layer is obtained. If the curing process is relatively slow as compared to that in "LC droplet-like morphology", transverse polymer filaments are obtained in polymer-rich region, in between

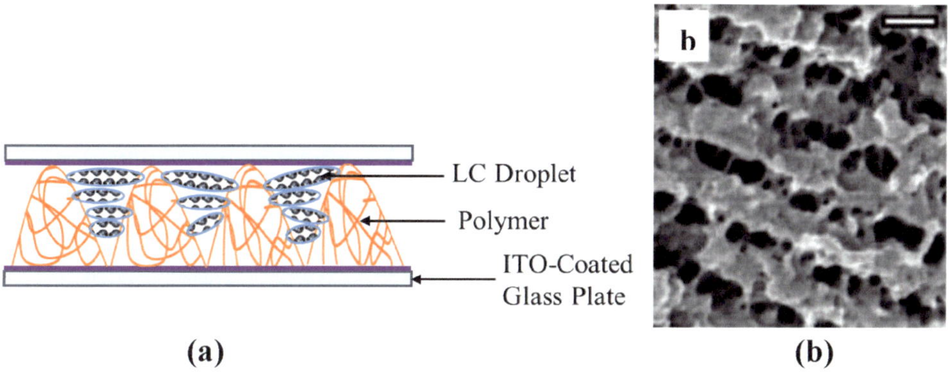

Figure 48.
LC droplet-like morphology: (a) schematic and (b) SEM image of HPDLC film [164, 168].

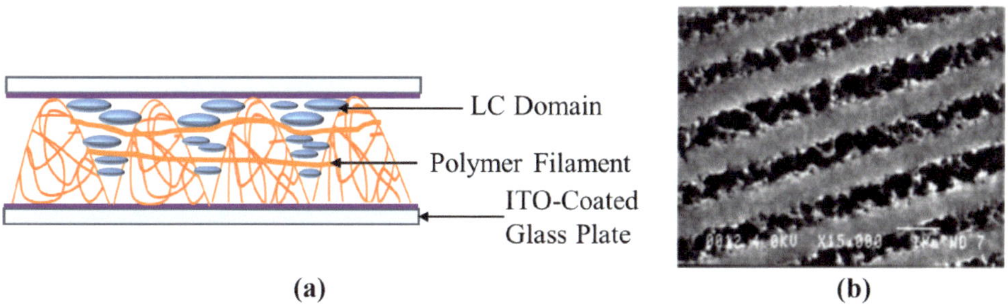

Figure 49.
Polymer scaffolding morphology: (a) schematic and (b) SEM image of HPDLC film [164, 169].

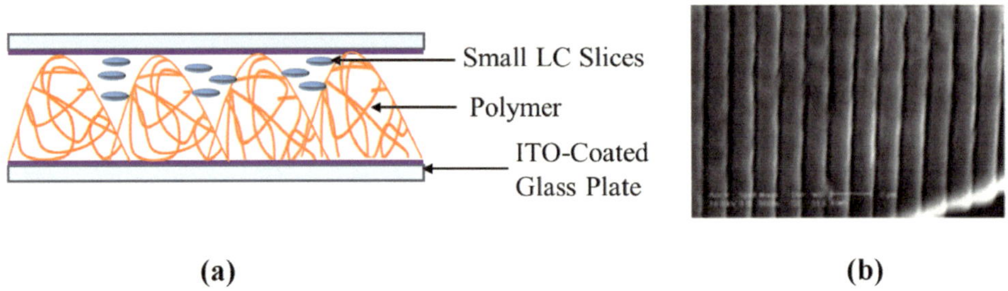

Figure 50.
Sliced polymer morphology: (a) schematic and (b) SEM image of HPDLC film [164, 170].

two LC domains/layers, as shown in SEM image (**Figure 49(b)**). As LC droplets are absent, scattering losses are low [164, 169].

7.5.3 Sliced polymer morphology (policryps)

In this type of morphology, transmission gratings are prepared from thiol-ene-based monomers, under slow curing process and above nematic-isotropic temperature T_{NI} of LC (**Figure 50**). In this type of grating, polymer slices are well separated from aligned nematic slices. From the SEM image (**Figure 50(b)**), it is clear that as the concentration of LC is low than the polymer, thickness of the LC slices is also small as compared to polymer slices. The high phase separation degree produces smooth polymer layers and aligned LC layers, which in turn minimizes scattering losses [164, 170].

7.6 Conclusions of HPDLC study

Similar to the PDLC films, HPDLC films are based on the scattering and transmittance effect of light from polymer-LC composite film. High monomer concentration and formation of interference pattern during photopolymerization give rise to alternate polymer-rich and LC-rich regions. Depending upon writing set-ups, two types of HPDLC gratings can be formed. Variation in materials and curing conditions can produce different types of morphology in HPDLC composite films. Simple configuration, easy fabrication process and their integration with other optical devices make them suitable for practical applications.

8. Applications

Polymer-LC composite film-based devices are switchable and tunable. The RI of LC droplets embedded in polymer matrix can be tuned using external fields; therefore they can be used for a wide range of display and non-display applications. The low monomer concentration polymer-LC composite film termed as PSLC films can be used for bi-stable reflective displays and haze-free normal- and reverse-mode light shutters with quick response [68, 110].

The PDLC composite films with the intermediate monomer concentration are distinguished because of their flexibility as well as mechanical strength. The comprehensive list of applications of PDLC film includes haze-free light shutter devices; switchable windows; high-definition Fuoss-Kirkwood spatial light modulators; flat-panel and large area flexible displays; light valves; color projectors; thermal, optical and strain sensors; electrically tunable focusing lenses; etc. [171].

When the monomer concentration is higher than the LC concentration, HPDLC films get formed. HPDLC films are known to be a promising technology because they can be used in 3D display, fiber optics, data storage, zoom lenses, image capture systems, beam vibration sensor, etc. [172].

It can be concluded that the LC's inherent optical and dielectric anisotropy can be effectively used in a display as well as non-display devices.

9. Summary

This chapter encompasses a wide range of LC-related subject matter. It begins with the basics of LC materials, such as definition, history, types, phases and properties of LCs. Information about properties of LC materials offer deep insight in choosing LC material for particular application and also benefit in innovating new applications. Initially applications of LC materials are limited to display world only; now it is finding scope in a non-display world too. A brief outline about applications of LC devices has been followed by an extensive study of polymer-LC composites. The three types of composite films reported in this chapter are as follows:

1. Low monomer concentration composite: PSLC film

2. Moderate monomer concentration composite: PDLC film

3. High monomer concentration composite: HPDLC film

The detailed study of each of the above composite films is about the following:

1. Fabrication technique using the most suited phase separation process.

2. Morphological analysis, which portrays their size, shape, configuration and defects in LC droplet/domain structure.

3. Electro-optic study: It comprises definition and formulation of various electro-optic parameters, such as transmittance difference, contrast ratio, hysteresis, threshold and saturation voltages and response time, with examples.

4. Dielectric study: Calculation of relaxation frequency, distribution parameter and breakdown strength using Debye and Cole-Cole plots is incorporated in this study.

Precisely, this chapter gives a deep and comprehensive knowledge about LC and some of their applications.

Author details

Anuja Katariya Jain and Rajendra R. Deshmukh*
Department of Physics, Institute of Chemical Technology, Mumbai, India

*Address all correspondence to: rr.deshmukh@ictmumbai.edu.in;
rajedeshmukh@rediffmail.com

References

[1] Woltman SJ, Crawford GP, Jay GD. Liquid Crystals: Frontiers in Biomedical Applications. Singapore: World Scientific; 2007

[2] Chandrasekhar S. Liquid Crystals. Cambridge: Cambridge University Press; 1992

[3] Reinitzer F. Zur Kenntnis des Cholesterins. Monatshefte für Chemie. 1888;**9**:421-441

[4] Dierking I. Textures of Liquid Crystals. Wiley-VCH Verlag: Weinheim; 2003

[5] Lehmann O. Uber fliessende Krystalle. Zeitschrift für physikalische Chemie. 1889;**4**:462-472

[6] Friedel G. The mesomorphic states of matter. Annales de Physique. 1922;**18**: 273-474

[7] Vorlander D. Chemische Kristallographie der Flussigkeiten: Kurze Anleitung zur Synthese und Untersuchung Polymorpher und Kristallinflüssiger Substanzen. Leipzig: Akademische Verlagsgesellschaft; 1924

[8] Herrmann K. Inclinations of molecules in some crystalline-fluid substances. Transactions of the Faraday Society. 1933;**29**:972-976

[9] Demus D, Goodby J, Gray GW, Spiess HW, Vill V. Handbook of Liquid Crystals-I. Wiley-VCH Verlag: Weinheim; 1998

[10] Jeu WHD. On the role of spherical symmetry in the Maier-Saupe theory. In: Cladis P, Palffy-Muhoray P, editors. Dynamics and Defects in Liquid Crystals: A Festschrift in Honor of Alfred Saupe. The Netherlands: Gordan and Breach; 1998

[11] Maier Z, Saupe A. Eine einfache molekular-statistische Theorie der nematischen kristallinflüssigen phase. Teil II. Zeitschrift für Naturforschung A. 1960;**15**:287-292

[12] Coates D, Gray GW. Optical studies of the amorphous liquid-cholesteric liquid crystal transition: The "blue phase". Physics Letters A. 1973;**45**: 115-116

[13] Demus D, Richter L, Rurup CE, Sackmann H, Schubert H. Synthesis and liquid crystalline properties of 4,4′-disubstituted biphenyls. Le Journal de Physique Colloques. 1975;**36**:349-354

[14] Luckhurst GR, Gray GW. The Molecular Physics of Liquid Crystal. London: Academic Press; 1979

[15] Andrienko D. Introduction to Liquid Crystals [Online]. 2006. Available from: www.mpip-ainz.mpg [Accessed: 11 July 2015]

[16] Demus D, Sackmann H. The problems of polymorphism in Liquid crystals. Molecular Crystals and Liquid Crystals. 1988;**21**:239-273

[17] LiqCryst Database. June 2010. Available from: www.lcipublisher.com/liqcryst.html [Accessed: 14 February 2019]

[18] Madsen LA, Dingemans TJ, Nakata M, Samulski ET. Thermotropic biaxial nematic liquid crystals. Physical Review Letters. 2004;**92**:145505-145514

[19] Drzaic PS. Liquid Crystal Dispersions. World Scientific: Singapore; 1995

[20] Collings PJ, Hird M. Introduction to Liquid Crystals. Bristol: Taylor & Francis; 1997

[21] De Gennes PG, Prost J. The Physics of Liquid Crystals. Oxford: Clarendon Press; 1993

[22] Bahadur B, editor. Liquid Crystals: Applications and Uses. Vol. I. Singapore: World Scientific; 1990

[23] Armitage D, Price FP. Calorimetry of liquid crystal phase transitions. Le Journal de Physique Colloques. 1975;**36**: 133-136

[24] Seideman T. The liquid-crystalline blue phases. Reports on Progress in Physics. 1990;**53**:659-705

[25] Crooker PP. Blue phases. In: Kitzerow HS, Bahr C, editors. Chirality in Liquid Crystals Partially Ordered System. New York: Springer; 2001. pp. 375-432

[26] Papon P, Lebond J, Meijer PHE. The Physics of Phase Transitions. Berlin Heidelberg: Springer-Verlag; 2006

[27] Largerwell ST. Comments on liquid crystal terminology, nomenclature and conventions. Ferroelectrics. 1988;**85**: 497-501

[28] Meiboom S, Sethna JP, Anderson PW, Brinkman WF. Theory of blue phase of cholesteric liquid crystals. Physical Review Letters. 1981;**46**: 1216-1219

[29] Brooks JD, Taylor GH. The formation of graphitizing carbons from the liquid phase. Carbon. 1965;**3**: 185-199

[30] Chandrasekhar S, Sadashiv BK, Suresh KA. Liquid crystals of disc like molecules. Pramana. 1977;**9**:471-480

[31] Desimpel, C. Liquid Crystals & Photonics Group, 2006. Available from: https://lcp.elis.ugent.be/tutorials/lc [Accessed: 01 December 2015]

[32] Pelzl G, Diele S, Weissflog W. Banana shaped compounds—A new field of liquid crystals. Advanced Materials. 1999;**11**:707-714

[33] Petrov AG. The Lyotropic State of Matter. Gordon & Breach: New York; 1999

[34] Stegemayer H. Liquid Crystals. New York: Springer; 1994

[35] Blinov LM, Cigrinov VG. Electro-Optic Effects in Liquid Crystal Materials. New York: SpringerVerlag; 1994

[36] Collings PJ. Liquid Crystals: Nature's Delicate Phase of Matter. Princeton: Princeton University Press; 1990

[37] Kedzierski J, Garbat K, Raszewski Z, Kojdecki MA, Kowiorski K, Jaroszewicz LR, et al. Optical properties of a liquid crystal with small ordinary and extraordinary refractive indices and small optical anisotropy. Opto-Electronics Review. 2014;**22**:162-165

[38] Mckenna L, Miller LS, Peterson IR. Polymer dispersed liquid crystal films for modulating infra-red radiation. Polymer. 2004;**45**:6977-6984

[39] Pasechnik SV, Chigrinov VG, Shmeliova DV. Liquid Crystals: Viscous and Elastic Properties. Wiley-VCH Verlag: Weinheim; 2009

[40] Singh S. Liquid Crystals Fundamentals. Singapore: World Scientific; 2002

[41] Taherian R. Application of conducting composite in dielectrics. In: Reza Taherian R, Kausar A, editors. Electrical Conductivity in Polymer-Based Composites: Experiments, Modelling and Applications. Oxford: William Andrew, Applied Science Publishers; 2019. pp. 73-90

[42] Leslie FM. Continuum theory for nematic liquid crystals. Continuum Mechanics and Thermodynamics. 1992; **4**:167-175

[43] Ge Z, Wu ST. Transflective Liquid Crystal Display. Chichester, UK: Wiley; 2010

[44] Kang H, Park JS, Sohn EH, Kang D, Rosenblatt C, Lee JC. Polyimide blend alignment layers for control of liquid crystal pretilt angle through baking. Polymer. 2009;**50**:5220-5227

[45] Mauguin MC. Sur les cristaux liquides de Lehmann. Bulletin de la Société Française de Minéralogie et de Cristallographie. 1911;**34**:71-117

[46] Varghese S. Patterned alignment of liquid crystal by microrubbing-A new route towards wide viewing angle flat panel displays [PhD thesis]. The Eindhoven University of Technology; 2005

[47] Shibaev VP. Polymers as Electrooptical and Photooptical Active Media. Heidelberg, Berlin: Springer-Verlag; 1996

[48] Acharya BR, Kim JH, Kumar S. Material independent determination of anchoring properties on rubbed polyimide surfaces. Physical Review E. 1999;**60**:6841-6846

[49] Janning JL. Thin film surface orientation for liquid crystals. Applied Physics Letters. 1972;**21**:173-174

[50] Porte G. Tilted alignment of MBBA induced by short-chain surfactants. Journal of Physics (France). 1976;**37**: 1245-1252

[51] Kim SH. Fast switching polymer stabilized liquid crystal devices: Morphological and electro-optical properties [PhD thesis]. Kent State University; 2004

[52] Van-Aerle NAJM, Toi AJW. Molecular reorientation in rubbed polyimide alignment layers used for liquid crystal display. Macromolecules. 1994;**27**:6520-6526

[53] Jeng SC, Hwang SJ. Controlling the alignment of polyimide for liquid crystal devices. In: Abadie MJM, editor. High Performance Polymers—Polyimides Based—from Chemistry to Applications. Rijeka: IntechOpen; 2012

[54] Yoon HG, Kang SW, Lehmann M, Park JO, Srinivasrao M, Kumar S. Homogenous and homeotropic alignment of bent core uniaxial and biaxial nematic liquid crystals. Soft Matter. 2011;7:8770-8775

[55] Kim JW, Kim H, Lee M, Magda JJ. Interfacial tension of a nematic liquid crystal/water Interface with homeotropic surface alignment. Langmuir. 2004;**20**:8110-8113

[56] Paek SH. Comparative study of effects of rubbing parameters on polyimide alignment layers and liquid crystal alignment. Journal of Industrial and Engineering Chemistry. 2001;**7**: 316-325

[57] Sonin AA. The Surface Physics of Liquid Crystals. Gordon and Breach: Luxembourg; 1995

[58] Bahadur B. Liquid crystals display. Molecular Crystals and Liquid Crystals. 1984;**109**:3-93

[59] Guyon, E., Urbach W. Anchoring properties and alignment of liquid crystals Kmetz, AR, von Willisen FK, editors. Non-Emissive Electro-optic Displays. Plenum Press: New York, 1976.

[60] Khoo IC. Liquid Crystals. Wiley: New Jersey; 2007

[61] Yang DK, Wu ST. Fundamentals of Liquid Crystal Devices. Chichester, UK: Wiley; 2006

[62] Lueder E. Liquid Crystal Display: Addressing Schemes and Electro-Optical Effects. Chichester, UK: Wiley; 2010

[63] Maini AK, Agrawal V. Electronic Devices and Circuits. New Delhi: Wiley; 2009

[64] Fergason JL. Display devices utilizing liquid crystal light modulation. US Patent 3731986, 1971.

[65] Helfrich W, Schadt M. Optical device. Swiss Patent 532261. 1970.

[66] Heilmeier GH, Zanoni LA. Guest-host interactions in nematic liquid crystals. Applied Physics Letters. 1968; **13**:91-92

[67] Scheffer TJ, Nehring J. Guets-host displays. In: The Physics and Chemistry of Liquid Crystal Devices. Springer: New York; 1980. pp. 173-180

[68] Crawford GP, Svensek D, Zumer S. Some aspects of polymer dispersed and polymer stabilized chiral liquid crystals. In: Kitzerow HS, Bahr C, editors. Chirality in Liquid Crystals Partially Ordered System. New York: Springer; 2001. pp. 375-432

[69] Jeon YJ, Bingzhu Y, Rhee JT, Cheung DL, Jamil M. Applications and new developments in polymer-dispersed liquid crystal simulation studies. Macromolecular Theory and Simulations. 2007;**16**:643-659

[70] Karapinar R. Electro-optic response of a polymer dispersed liquid crystal film. The Turkish Journal of Physics. 1998;**22**:227-235

[71] Spruce G, Pringle RD. Polymer dispersed liquid crystal (PDLC) films. Electronics & Communication Engineering Journal. 1992;**4**:91-100

[72] Liu YJ, Sun XW. Holographic polymer-dispersed liquid crystals: Materials, formation, and applications. Advances in OptoElectronics. 2008; **2008**:1-52

[73] Crawford GP, Zumer S. Liquid Crystals in Complex Geometries. London: Taylor & Francis; 1996

[74] Hicks SE. Polymer dispersed and polymer stabilized liquid crystals [PhD thesis]. Kent State University; 2012

[75] Zou J, Fang J. Director configuration of liquid-crystal droplets encapsulated by polyelectrolytes. Langmuir. 2010;**26**: 7025-7028

[76] Crawford NJ, Dadmun MD, Bunning TJ, Natarajan LV. Time-resolved light scattering of the phase separation in polymer-dispersed liquid crystals formed by photo-polymerization induced phase separation. Polymer. 2006;**47**:6311-6321

[77] Dierking I. Polymer network stabilized liquid crystals. Advanced Materials. 2002;**12**:167-181

[78] Fung YK, Yang DK, Ying S, Chien LC, Zumer S, Doane JW. Polymer networks formed in liquid crystals. Liquid Crystals. 1995;**19**:797-801

[79] Kumar R, Raina KK. Enhanced ordering in polymer stabilised ferroelectric liquid crystal guest-host composites: Evidence by polarised fluorescence spectroscopy. Liquid Crystals. 2014;**41**:694-700

[80] Deshmukh RR, Katariya-Jain A. Effect of anti-parallel and twisted alignment techniques on various properties of polymer stabilized Liquid crystal (PSLC) films. Liquid Crystals. 2016;**43**:436-447

[81] Moheghi A, Nemati H, Yang DK. Polarizing light waveguide plate from polymer stabilized liquid crystals. Optical Materials Express. 2015;**5**: 217-1223

[82] Serra F, Buscaglia M, Bellini T. The emergence of memory in liquid crystals. Materials Today. 2011;**14**:488-490

[83] Malik MK, Deshmukh RR. Electro-optics of homogenously aligned Nematic Liquid crystals stabilized by a

polymer network. International Journal of ChemTech Research. 2014;**6**:1833-1835

[84] Deshmukh RR, Malik MK. Effects of the composition and nematic-isotropic phase transition on the electro-optical responses of unaligned polymer-dispersed liquid crystals. I. Composites of poly(methyl methacrylate) and E8. The Journal of Applied Polymer Science. 2008;**108**:3063-3072

[85] Dixit S. Liquid crystals: Dielectric and electro-optic study. [PhD thesis]. Lucknow: University of Lucknow; 2013

[86] Parab SS, Malik MM, Deshmukh RR. Dielectric relaxation and electro-optical switching behavior of nematic liquid crystal dispersed in poly(methyl methacrylate). Journal of Non-Crystalline Solids. 2012;**358**:2713-2722

[87] Thakur VK, Kessler MR. Polymer nanocomposites: New advanced dielectric materials for energy storage applications. In: Tiwari A, Valyukh S, editors. Advance Material Energy. Salem, MA: Scrivener Publishing LLC-Wiley; 2014

[88] Dissado LA, Fothergill LC. Electrical Degradation and Breakdown in Polymers. London, UK: Peter Peregrins; 1992

[89] Malik NH, Al-Arainy AA, Qureshi MI. Electrical Insulation in Power System. Boca Raton: CRC Press; 1998

[90] Bunget I, Popescu M. Physics of Solid Dielectrics. Amsterdam, The Netherlands: Elsevier; 1984

[91] Raju GG. Dielectrics in Electric Fields. New York: CRC Press; 2003

[92] Jonscher AK. Dielectric Relaxation in Solids. Chelsea Dielectric Press: London; 1983

[93] Ding H. A study of dielectric and electro-optical response of liquid crystal in confined systems. [PhD thesis]. Kent State University; 1996

[94] Seekola DL. Dielectric response of polymer dispersed liquid crystal films [PhD thesis]. Kent State University; 1992

[95] Deshmukh RR, Parab SS, Malik MK. Effect of host polymer matrices on electro optical and dielectric behaviour of polymer dispersed liquid crystal system. Advances in Materials Research. 2012;**584**:531-535

[96] Lee DG, Suh NM. Axiomatic Design and Fabrication of Composite Structures: Applications in Robots, Machine Tools, and Automobiles. New York: Oxford University Press; 2006

[97] Kremer F, Schonhals A. Broadband Dielectric Spectroscopy. Berlin Heidelberg: Springer-Verlag; 2003

[98] Wrobel S. Dielectric relaxation spectroscopy. In: Haase W, Wrobel S, editors. Relaxation Phenomena: Liquid Crystals, Magnetic Systems, Polymers, High-Tc Superconductors, Metallic Glasses. Berlin, Heidelberg: Springer-Verlag; 2003

[99] Lvovich VF. Impedance Spectroscopy Applications to Electrochemical and Dielectric Phenomena. Wiley: New Jersey; 2012

[100] Parab SS, Malik MK, Bhatia PG, Deshmukh RR. Investigation of liquid crystal dispersion and dielectric relaxation behavior in polymer dispersed liquid crystal composite films. Journal of Molecular Liquids. 2014;**199**:287-293

[101] Parab SS, Malik MM, Deshmukh RR. Investigation of dielectric properties of poly(methyl methacrylate)-E7 composite films. International Journal of ChemTech Research. 2014;**6**:1836-1839

[102] Cole KS, Cole RH. Dispersion and absorption in dielectrics-I alternating

current characteristics. The Journal of Chemical Physics. 1941;**9**:341-351

[103] Cole KS, Cole RH. Dispersion and absorption in dielectrics-I alternating current characteristics. The Journal of Chemical Physics. 1942;**10**:98-105

[104] Al-Refaie SN. A generalized formula to determine the relaxation time distributions in dielectrics. Applied Physics A: Materials Science & Processing. 1991;**52**:234-236

[105] Parab SS, Malik MK, Dabrowski R, Deshmukh RR. Thermodynamic and bias field characterization of quickly operating antiferroelectric liquid crystal. Journal of Molecular Liquids. 2013;**183**: 20-25

[106] Pevzner B. Transport and dielectric properties of thin fullerene (C60) films [PhD thesis]. Massachusetts: Massachusetts Institute of Technology; 1995

[107] Gray GW, Kelly SM. Liquid crystals for twisted nematic display devices. Journal of Materials Chemistry. 1999;**9**:2037-2050

[108] Hale PS, Shapter JG, Voelcker NH, Ford MJ, Waclawik ER. Liquid-crystal displays: Fabrication and measurement of a twisted nematic liquid-crystal cell. Journal of Chemical Education. 2004;**81**: 854-860

[109] Rajaram CV, Hudson SD. Morphology of polymer stabilized Liquid crystal. Chemistry of Materials. 1995;**7**:2300-2308

[110] Zhang T, Cong Y, Zhang B, Zhao W. Preparation and characterisation: PSCLC film doping with FexNiy nanoparticles. Liquid Crystals. 2015;**42**: 167-173

[111] Sun H, Xie Z, Ju C, Hu X, Yuan D, Zhao W, et al. Dye-doped electrically smart windows based on polymer-stabilized liquid crystal. Polymer. 2019;**11**: 694-703. DOI: 10.3390/polym11040694

[112] Yan X, Zhou Y, Liu W, Liu S, Hu X, Zhao W, et al. Effects of silver nanoparticle doping on the electro-optical properties of polymer stabilized liquid crystal devices. Liquid Crystals. 2019;**46**:1-8. DOI: 10.1080/02678292.2019.1641754

[113] Fergason JL. Encapsulated liquid crystal and method. US Patent 4435047, 1984.

[114] Fergason JL. Polymer encapsulated nematic liquid crystals for display and light control applications. Society for Information Display Technical Digest. 1985;**85**:68-70

[115] Doane JW, Golemme A, West JL, Whitehead JB Jr, Wu BG. Polymer dispersed liquid crystals for display application. Molecular Crystals and Liquid Crystals. 1988;**165**:511-532

[116] Deshmukh RR. Electro-optic and Dielectric properties in PDLC composite systems. In: Thakur VK, Kessler MR, editors. Liquid Crystalline Polymers-II, Processing and Applications. Switzerland: Springer International Publishing; 2015. pp. 169-195

[117] Kalkar AK, Kunte VV. Electro-optical studies on polymer dispersed liquid crystal composite films. II. Composites of PVB/E44 and PMMABA/E44. Molecular Crystals and Liquid Crystals. 2002;**383**:1-25

[118] Li W, Cao Y, Cao H, Kashima M, Kong L, Yang H. Effects of the structures of polymerizable monomers on the electro-optical properties of UV cured polymer dispersed liquid crystal films. Journal of Polymer Science Part B: Polymer Physics. 2008;**46**:1369-1375

[119] Liu F, Cao H, Mao Q, Song P, Yang H. Effects of monomer structure on the morphology of polymer networks and

the electro-optical properties of polymer-dispersed liquid crystal films. Liquid Crystals. 2012;**39**:419-424

[120] Lu Y, Wei J, Shi Y, Jin O, Guo J. Effects of fabrication condition on the network morphology and electro-optical characteristics of polymer-dispersed bistable smectic a liquid crystal device. Liquid Crystals. 2013;**40**:581-588

[121] West JL, Ondris R, Erdmann M. Dichroic dye containing polymer-dispersed liquid crystal films. Proceedings of SPIE. 1990;**1257**:76-83

[122] Doane JW, Vaz NA, Wu BG, Zumer S. Field controlled light scattering from nematic microdroplets. Applied Physics Letters. 1986;**48**: 269-271

[123] Bulgakova SA, Mashin AI, Kazantseva IA, Kashtanov DE, Jones MM, Tsepkov GS, et al. Influence of the composition of the polymer matrix on the Electrooptical properties of films with a dispersed liquid crystal. Russian Journal of Applied Chemistry. 2008;**81**: 1446-1451

[124] Higgins DA. Probing the mesoscopic chemical and physical properties of polymer-dispersed liquid crystals. Advanced Materials. 2000;**12**: 251-264

[125] Kumar P. Preparation and characterization of dichroic polymer dispersed liquid crystals [PhD thesis]. Patiala: Thapar University; 2007

[126] Frank FC. Liquid crystals. On the theory of liquid crystals. Transactions of the Faraday Society. 1958;**25**:19-28

[127] Drzaic PS. a new director alignme nt for droplets of nematic liquid crystal with low bend-to-splay ratio. Molecular Crystals and Liquid Crystals. 1988;**154**: 289-306

[128] Erdmann JH, Zumer S, Doane JW. Configuration transition in a nematic liquid crystal confined to a small spherical cavity. Physical Review Letters. 1990;**64**:1907-1910

[129] Golemme A, Zumer S, Allender DW, Doane JW. Continuous nematic-isotropic transition in submicron-size liquid-crystal droplets. Physical Review Letters. 1988;**61**:2937-2940

[130] Volovik GE, Lavrentovich OD. Topological dynamics of defects: Boojums in nematic drops. Soviet Physics—Journal of Experimental and Theoretical Physics. 1983;**58**:1159-1166

[131] Brinkman WF, Cladis PE. Defects in liquid crystals. Physics Today. 1982; **35**:48-54

[132] Candau S, Le Roy P, Debeauvais F. Magnetic field effects in nematic and cholesteric droplets suspended in an isotropic liquid. Molecular Crystals and Liquid Crystals. 1973;**23**:283-297

[133] He J, Yan B, Yu B, Wang S, Zeng Y, Wang Y. The effect of molecular weight of polymer matrix on properties of polymer-dispersed liquid crystals. European Polymer Journal. 2007;**43**: 2745-2749

[134] Deshmukh RR, Katariya-Jain A. Novel techniques of PDLC film preparation furnishing manifold properties in a single device. Liquid Crystals. 2016;**43**:256-267

[135] Malik MK, Bhatia PG, Deshmukh RR. Effect of nematic liquid crystals on optical properties of solvent induced phase separated PDLC composite films. Indian Journal of Science and Technology. 2012;**5**:3440-3452

[136] Kalkar AK, Kunte VV, Bhamare SA. Electrooptic studies on polymer-dispersed liquid-crystal composite films. III. Poly(methyl methacrylate-Co-butyl acrylate)/E7 and poly(methyl methacrylate-Co-butyl acrylate)/E8 composites. The Journal of Applied Polymer Science. 2008;**107**:689-699

[137] Wu BG, West JL, Doane JW. Angular discrimination of light transmission through polymer-dispersed-liquid-crystal films. Journal of Applied Physiology. 1987;**62**:3925-3931

[138] Drzaic PS. Reorientation dynamics of polymer dispersed nematic liquid crystal films. Liquid Crystals. 1988;**3**: 1543-1559

[139] Wu BG, Erdmann JH, Doane JW. Response times and voltages for PDLC light shutters. Liquid Crystals. 1989;**5**: 1453-1465

[140] Deshmukh RR, Malik MK. Effect of temperature on the optical and electro-optical properties of poly (methyl methacrylate)/E7 polymer-dispersed Liquid crystal composites. Journal of Applied Polymer Science. 2008;**109**:627-637

[141] Reamey RH, Montoya W, Wong A. Video microscopy of NCAP films: The observation of LC droplet in real time. Proceedings of SPIE. 1992;**1665**:2-7

[142] Hirai Y, Niiyama S, Ooi Y, Kumai H, Wakabayashi T, Gunjima T. Recent development in polymer dispersed LCDS. Society for Information Display Technical Digest. 1994;**25**:833-836

[143] Deshmukh RR, Katariya-Jain A. The complete morphological, electro-optical and dielectric study of dichroic dye doped polymer dispersed liquid crystal. Liquid Crystals. 2014;**41**: 960-975

[144] Han JW, Kang TJ, Park G. Effects of composition, curing-time, and temperature on the electro-optical characteristics of polymer-dispersed liquid crystal films. Journal of the Korean Physical Society. 2000;**36**: 156-163

[145] Neeraj, Raina KK. Multiwall carbon nanotubes doped ferroelectric liquid crystal composites: A study of modified electrical behavior. Physica B. 2014; **434**:1-6

[146] Sadovoy AV, Nazvanov VF. Optical transmission of polymer dispersed liquid crystals doped with carbon nanotubes. Physics Letters. 2006;**32**:659-660

[147] Song P, Cao H, Wang F, Liu F, Yang H. The influence of the structure of curable epoxy monomers on the electro-optical properties of polymer dispersed liquid crystal devices prepared by UV initiated cationic polymerization. Liquid Crystals. 2012; **39**:433-440

[148] Wang JH, Zhang BY, Qu WZ, Chu HS, Li H. The kinetics of photo-polymerization in the fabrication of polymer-dispersed liquid crystals doped with nano-graphite. Liquid Crystals. 2010;**37**:1-11

[149] Yadav SP, Pandey KK, Misra AK, Dixit S, Manohar R. Molecular dynamics in weakly polar nematic liquid crystal doped with dye. Canadian Journal of Physics. 2011;**89**:661-665

[150] Bauman D, Hasse W. Dielectric measurements of guest-host systems. Molecular Crystals and Liquid Crystals. 1989;**168**:155-168

[151] Drzaic PS. Recent progress in dichroic polymer-dispersed liquid crystal materials. Pure and Applied Chemistry. 1996;**68**:1435-1440

[152] Malik P, Raina KK. Dichoic dye - dependent studies in guest-host polymer dispersed liquid crystal films. Physica B: Condensed Matter. 2010;**405**: 161-166

[153] Masutani A. Guest-host effect of dyes in polymer dispersed liquid crystals [PhD thesis]. Durham: Durham University; 2002

[154] Kumar P, Raina KK. Morphological and electro-optical responses of dichroic

polymer dispersed liquid crystal films. Current Applied Physics. 2007;**7**: 636-642

[155] Deshmukh RR, Malik MK. A method for estimating interfacial tension of liquid crystal embedded in polymer matrix forming PDLC. The Journal of Applied Polymer Science. 2014;**131**:41137 (1–6)

[156] Deshmukh RR, Malik MK. Effect of dichroic dye on phase separation kinetics and electro-optical characteristics of polymer dispersed liquid crystals. Journal of Physics and Chemistry of Solids. 2013;**74**:215-224

[157] Deshmukh RR, Malik MK. Photopolymerisation kinetics and electro-optical properties in mixtures of dichroic dye doped nematic liquid crystal and photocurable polymer. Liquid Crystals. 2013;**40**:1050-1059

[158] Deshmukh RR, Malik MK, Parab SS. Dichroic dye induced nonlinearity in polymer dispersed Liquid crystal materials for display devices. Advances in Materials Research. 2012;**584**:79-83

[159] Carlton RA. Pharmaceutical Microscopy. Springer: New York; 2011

[160] Katariya-Jain A, Deshmukh RR. Influence of a guest dichroic azo dye on a host liquid crystal dispersed in polymer matrix. International Journal of ChemTech Research. 2014;**6**:1813-1816

[161] Lagerwall JPF, Scalia G. Carbon nanotubes in liquid crystals. Journal of Materials Chemistry. 2008;**18**:2890-2898

[162] Katariya-Jain A, Deshmukh RR. Electro-optical and dielectric study of multi-walled carbon nanotube doped polymer dispersed liquid crystal films. Liquid Crystals. 2019;**46**:1191-1202

[163] Domash LH, Crawford GP, Ashmead AC, Smith RT, Popovich MM, Storey J. Holographic PDLC for photonic applications. Proceedings of SPIE. 2000;**4107**:46-58

[164] Ma J, Huang W, Xuan L, Yokoyama H. Holographic polymer-dispersed liquid crystals: From materials and morphologies to applications. In: Vaibhav Jain V, Kokil A, editors. Optical Properties of Functional Polymers and Nano Engineering Applications. Boca Raton: CRC Press, Taylor & Francis Group; 2015

[165] Villa-Manríquez JF, Ortiz-Gutiérrez M, Pérez-Cortés M, Ibarra-Torre JC, Arturo Olivares-Pérez M. Holographic gratings recorded in PDLC mixed with crystal violet dye. Optik. 2017;**144**:219-223

[166] Sutherland RL, Natarajan LV, Tondiglia VP. Bragg gratings in an acrylate polymer consisting of periodic polymer-dispersed liquid-crystal planes. Chemistry of Materials. 1993;**5**:1533-1538

[167] Wang K, Zheng J, Liu Y, Gao H, Zhuang S. Electrically tunable two-dimensional holographic polymer-dispersed liquid crystal grating with variable period. Optics Communication. 2017;**392**:128-134

[168] Jakubiak R, Bunning TJ, Vaia RA, Natarajan LV, Tondiglia VP. Electrically switchable, one-dimensional polymeric resonators from holographic photopolymerization: A new approach for active photonic Bandgap materials. Advanced Materials. 2003;**15**:241-244

[169] Sarkar MD, Gill NL, Whitehead JB, Crawford GP. Effect of monomer functionality on the morphology and performance of the holographic transmission gratings recorded on polymer dispersed liquid crystals. Macromolecules. 2003;**36**:630-638

[170] Caputo R, Veltri A, Umeton C, Sukhov AV. Kogelnik-like model for the

diffraction efficiency of POLICRYPS gratings. Journal of the Optical Society of America B: Optical Physics. 2005;**22**: 735-742

[171] Coates D. Polymer dispersed liquid crystals. Journal of Materials Chemistry. 1995;**5**:2063-2072

[172] Liu Y, Zheng J, Shen T, Wang K, Zhuang S. Diffusion kinetics investigations of nano Ag-doped holographic polymer dispersed liquid crystal gratings. Liquid Crystals. 2019; **46**:1050-1059

Chapter 3

Cholesteric Liquid Crystal Polyesteramides: Non-Viral Vectors

Mercedes Pérez Méndez and José Fayos Alcañiz

Abstract

Polyesteramides PNOBDME $(C_{34}H_{38}N_2O_6)_n$, Poly[oxy(1,2-dodecane)-oxy-carbonyl-1,4-phenylene-amine-carbonyl-1,4-phenylene-carbonyl-amine-1,4-phenylene-carbonyl], and PNOBEE $(C_{26}H_{22}N_2O_6)_n$, Poly[oxy(1,2-butylene)-oxy-carbonyl-1,4-phenylene-amine-carbonyl-1,4-phenylene-carbonyl-amine-1,4-phenylene-carbonyl], have been designed and synthesized as cholesteric liquid crystals (LCs)—through a condensation reaction between 4- 4′-(terephthaloyl-diaminedibenzoic chloride) (NOBC) and racemic glycol, DL-1,2-dodecanediol or DL-1,2-butanediol, respectively—as chemical modifications of multifunctional cholesteric LC polyesters, involving new properties but holding the precursor helical macromolecular structures. The new compounds have been characterized by ^{1}H and ^{13}C-NMR, COSY and HSQC, exhibiting two ^{1}H-independent sets of signals observed for each enantiomer, attributed to two diastereomeric conformers, *gg* and *gt*, of the torsion containing the asymmetric carbon atom in the spacer. They have also been characterized by x-ray diffraction with synchrotron radiation source. Thermal behaviour of the new compounds is studied by thermogravimetric (TG) and differential scanning calorimetry (DSC) analysis. The substitution of the ester groups in the mesogen by amide groups causes an increase of thermal stability with respect to the precursors. Optical rotatory dispersion (ORD) is evaluated. Morphology of powdered PNOBDME exhibits spherical clusters of about 5 μm in diameter homogeneously dispersed. Molecular models show helical polymeric chains with stereoregular head-tail, isotactic structure, explained as due to the higher reactivity of the primary hydroxyl with respect to the secondary one in the glycol through the polycondensation reaction. Besides being biocompatible, these synthetic polyesteramides have proved to act as non-viral vectors in gene therapy and be able to transfect DNA to the nucleus cell. Similar new cationic cholesteric liquid crystal polyesters have also been synthesized in our laboratory.

Keywords: cholesteric LC polymer, biocompatible polyesteramides, synthesis, characterization, SAXS/WAXS, molecular simulation

1. Introduction

In 1944, Erwin Schrödinger, one of the founders of quantum mechanics, completed in Dublin a small book entitled *What Is Life? The Physical Aspect of the Living Cell* [1]. He predicted there the concept of "aperiodic crystal", to define a gene or perhaps the whole fibre of the chromosome, "that in my opinion, is the material carrier of life...," "able to grow in aggregates bigger and bigger without the clumsy resource of repetition in three directions" as crystals do [2].

The concept of "aperiodic crystal" was surprisingly premonitory 9 years before the structure of DNA fibres was determined by x-ray diffraction as a helical structure by Franklin [3] ("Photograph 51" in **Figure 1(a)**), Wilkins [5] and Watson and Crick who reported the double helix with periodic order along the sugar-phosphate helical backbone, every 34 Å, and with the *aleatory* distribution of the complementary base pairs [4] (**Figure 1(b)**). The sequence of base pairs along the structure, without lateral periodic order, being convenient to endow DNA of their capacity to store genetic information.

In 1988 Ringsdorf evidenced the parallelism between thermotropic and lyotropic liquid crystals (LCs) in materials science and lipids in life science, with common amphiphilic nature [6]. Both self-organize their molecules in supramolecular systems, leading to functional units in highly oriented systems, exhibiting new properties. The importance of lyotropic LCs for the life sciences has been known for a long time as a prerequisite for the development of life and the ability of cells to function. In materials sciences, the concept of function through organization led to the development of new liquid crystalline materials for advanced applications.

Since 1992 the International Union of Crystallography redefined the concept of crystal as: "Any solid which has a diffraction pattern essentially discrete" (Fourier space) [7]. Since then, the *crystal family* was accepted to be composed of *periodic* and *aperiodic crystals*, and liquid crystals belong to the last group.

With increasing temperature, they do not directly go from the crystalline state into the melt, but, in the middle, they undergo a *mesophase* state which combines the order of perfect crystals and mobility of liquids.

All mesophases exhibit long-range orientational order, by keeping the parallel orientation of their longitudinal molecular axes. Two major classes can be

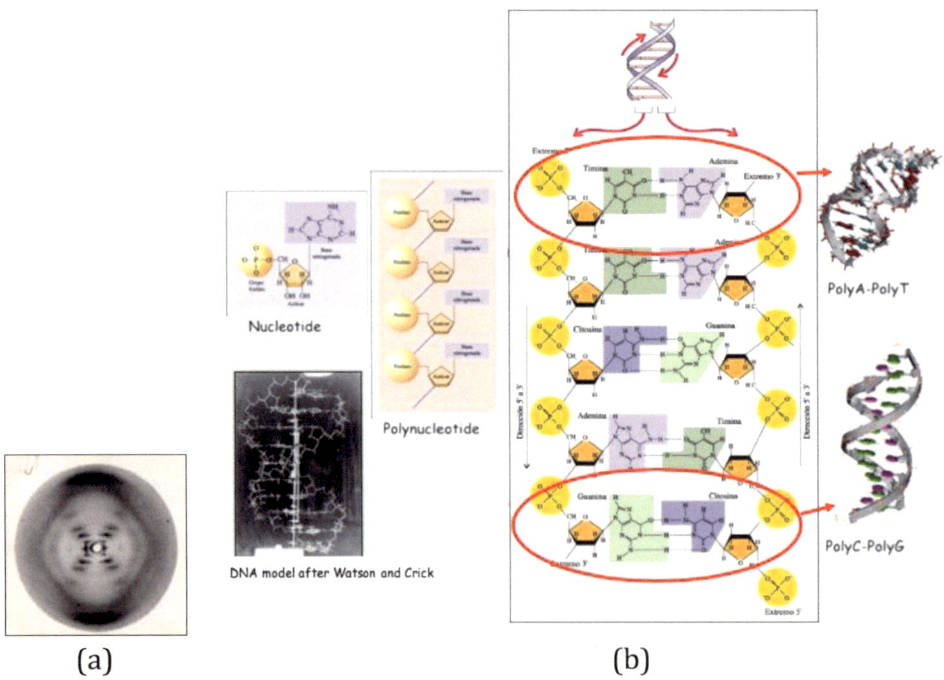

Figure 1.
(a) Photograph 51 by [3]; (b) DNA crystal structure according to [4]. Details of double-stranded PolyA-PolyT and PolyC-PolyG are shown at the right-hand side.

differentiated: *nematic* (with molecular centres distributed isotropically) and *smectic* (with their molecular centres organized in layers). The special array of nematic planes stacked in a helical superstructure with a prevalent screw direction is called the cholesteric mesophase (**Figure 2**).

DNA behaves as liquid crystal with the special array of the nematic planes, containing complementary base pairs, stacked in a superstructure with chiral helical symmetry of the charge distribution [8–10]. **Figure 3** shows micrographs of the cholesteric mesophase of DNA as the columnar hexagonal phase.

1.1 Synthetic cholesteric liquid crystal polymers

Two multifunctional cholesteric liquid crystal polyesters (ChLCP), named PTOBDME $[C_{34}H_{36}O_8]_n$ and PTOBEE $[C_{26}H_{20}O_8]_n$, with chemical formulations in **Figure 4**, were synthesized in our laboratory—through a condensation reaction between 4-4′-(terephthaloyl-dioxybenzoylchloride) (TOBC) and racemic glycol, DL-1,2-dodecanediol or DL-1,2-butanediol, respectively.

Although only racemic materials were used in their synthesis, a cholesteric, chiral morphology, theoretically unexpected, was found for PTOBDME, m = 9 in **Figure 4**. Evidence of this was obtained when a white solid, recrystallized as the second fraction from toluene mother liquor after the filtration of the polymer, was identified as –PTOBDME, with $[\alpha]^{25}_{589} = -1.43$ (1.538 g/100 ml, toluene). A similar result had been previously attained for liquid crystal PTOBEE, m = 1 in **Figure 4**. Its second fraction was isolated as –PTOBEE, with a value of $[\alpha]^{25}_{589} = -2.33$ (0.0056 mol/l, toluene). The synthetic method [11, 12], based on the previously reported by Bilibin [13, 14], leads to the obtaining of two or more fractions with progressively enriched diastereomeric excess. Its structure and diastereomeric excess could be characterized by NMR [15].

The structure of these optically active cholesteric LC polymers was characterized by NMR, Raman spectroscopy, steady-state fluorescence and SAXS/WAXS [16, 17], as rigid or semirigid helical LC polymers chains with flexible branches (chiral groups are located in the backbone), **Figure 5**, one of the major types of macromolecules forming the cholesteric mesophase [18].

Both polymers behave as thermotropic and lyotropic. They self-assemble in nanocavities in solution, with different conformations depending on the solvent

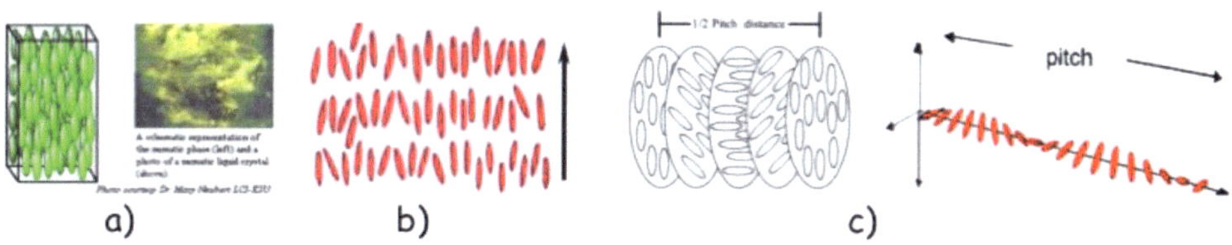

Figure 2.
Types of liquid crystal mesophases: (a) nematic, (b) smectic a and (c) cholesteric, helical pitch.

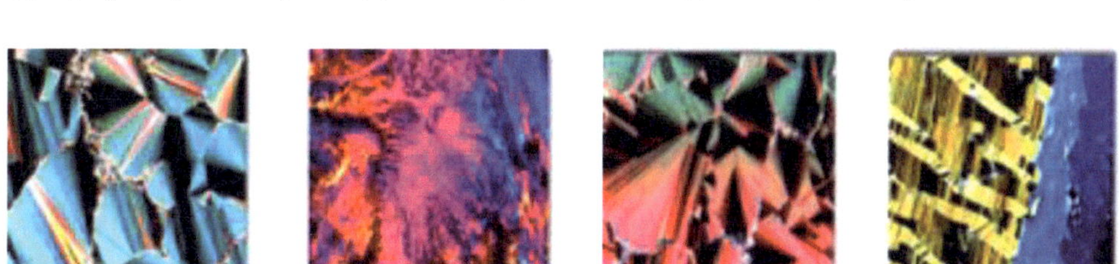

Figure 3.
Cholesteric mesophase of liquid crystal DNA exhibiting columnar hexagonal phase.

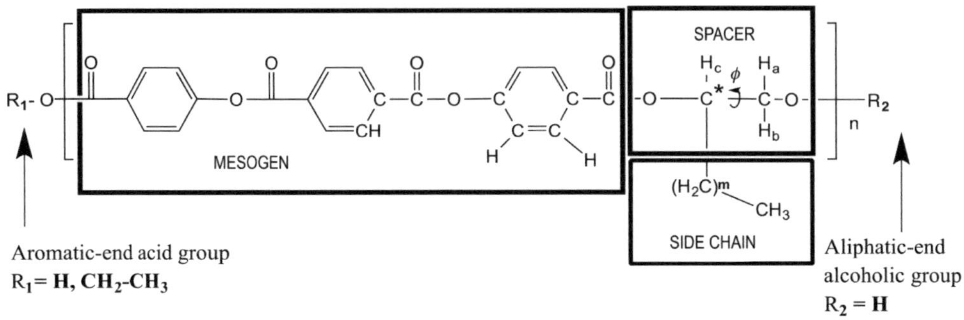

Figure 4.
The monomeric unit of cholesteric liquid crystal polyesters PTOBEE (m = 1) and PTOBDME (m = 9). The three different zones of the monomer are indicated: mesogen, spacer *and* flexible side chain. *The chiral Centre is indicated by an asterisk. Torsion ϕ is shown. The aromatic-end acid and aliphatic-end alcoholic groups are also shown.*

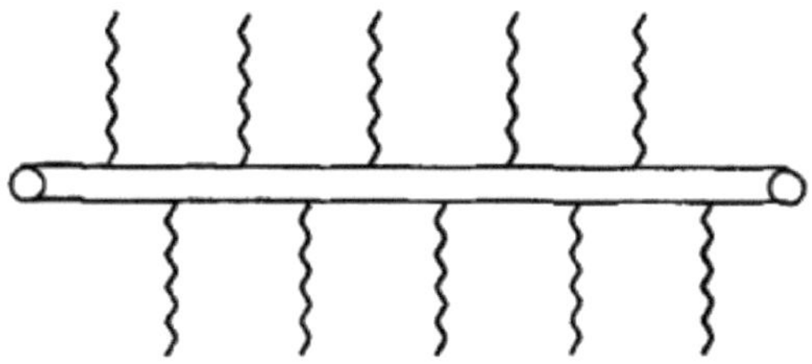

Figure 5.
Schematic representation of the cholesteric LC polymer type, with a rigid or semirigid helical chain with flexible branches (chiral groups are located in the backbone).

and on the concentration [16]. They also get adsorbed on metal surfaces with reordering of the polymer in the interface [19], with a potential application on the biomedical and engineering field.

Besides they have proved to be biocompatible against macrophages and fibroblasts cellular lines. They are also able to interact with biomacromolecules: lipids (neutral and cationic), polynucleotides and nucleic acids.

1.2 Interaction with lipids

Figure 6 shows liposomes of L-α-DMPC when interacting with cholesteric liquid crystal polyester PTOBDME, adopting hexagonal form. The structures of the lipid membranes complexed with cholesteric LC polymer were analysed by simultaneous SAXS/WAXS with synchrotron radiation source with $\lambda = 1.5$ Å on the X33

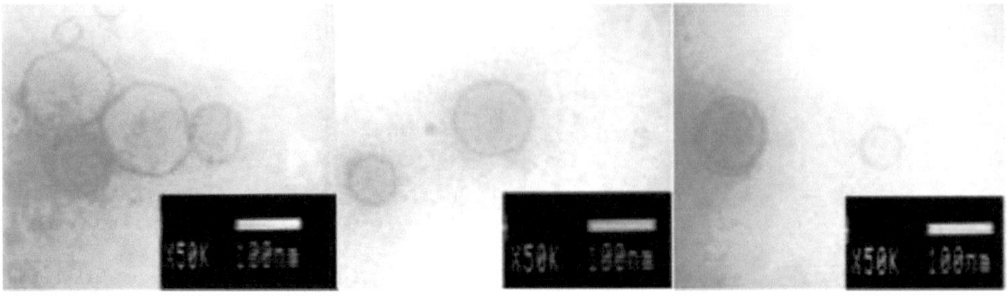

Figure 6.
TEM images of L-α-DMPC hexagonal liposomes complexed with PTOBDME.

camera at EMBL (DESY, Hamburg) [20–22]. Tripalmitin was used to calibrate both linear detectors. All data were normalized for incident intensity and analysed with Primus (ATSAS) [23, 24].

1.3 Interaction with polynucleotides and nucleic acids: non-viral vectors

The entrance of exogenous genetic material in cells was a key stage in the development of cellular biology. The term "transfection" indicates the transfer of DNA—as a healing agent—into the nuclei of cells of higher organisms. The direct application of this technology in living organisms opened crucial possibilities, like gene therapy and DNA vaccines [25].

Synthetic molecules that can bind polynucleotide fragments (the therapeutic agent) are required to develop new non-viral vectors to transfect in cells, without stimulating an immune response.

Cationic polymers, at physiological pH, are used to condense anionic nucleic acids, through self-assembly driven by electrostatic interactions, into nano-sized complexes called "polyplexes." DNA molecules being compressed to a relatively smaller size able to enter inside the nuclei of cells, facilitating internalization, thus improving transfection efficacy [26].

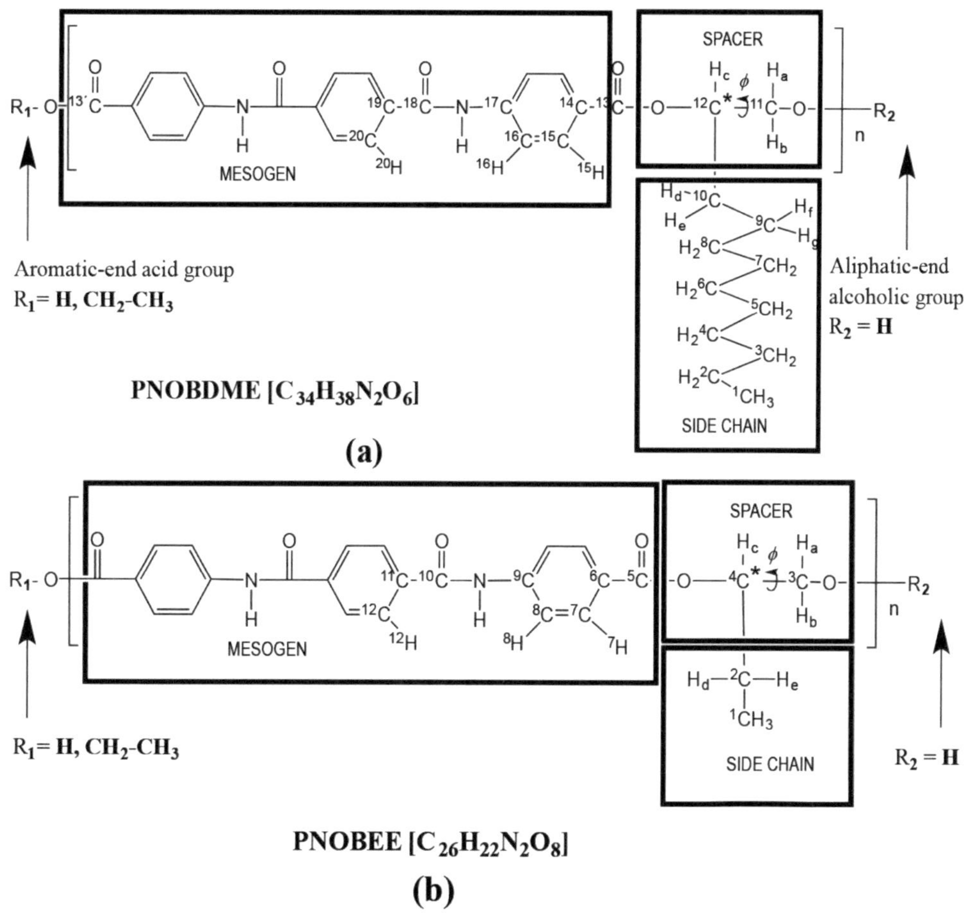

Figure 7.
The monomeric unit of polyesteramides: (a) PNOBDME; (b) PNOBEE. The asterisk indicates the chiral centre ($^{12}C^*$) in PNOBDME and ($^{4}C^*$) in PNOBEE. Torsion angle φ, along the ^{11}C–$^{12}C^*$ bond and ^{3}C–$^{4}C^*$ bond, respectively, is indicated.

New polyplexes were formulated including cholesteric LC PTOBDME and/or PTOBEE and *new monomeric cationic surfactant molecules* synthesized in our lab to entrap an anionic DNA plasmid. The new polyplexes successfully transfected both in vitro and in vivo in mice, as non-viral vectors for gene therapy. Their structures were studied by synchrotron radiation source [22, 27, 28].

New cationic chemical formulations of multifunctional cholesteric LCs PTOBDME and PTOBEE were designed later to directly interact with negatively charged DNA. The functional groups selected to be introduced at the end of the main polyester chains were choline [$-CH_2-CH_2-N^+-(CH_3)_3$] and ammonium, defined as [$-CH_2-CH_2-CH_2-N^+H-(CH_3)_2$]. Two more new polymers were also formulated with amide groups ($-CONH_2$) chemically bonded to the end of the lateral hydrophobic chains.

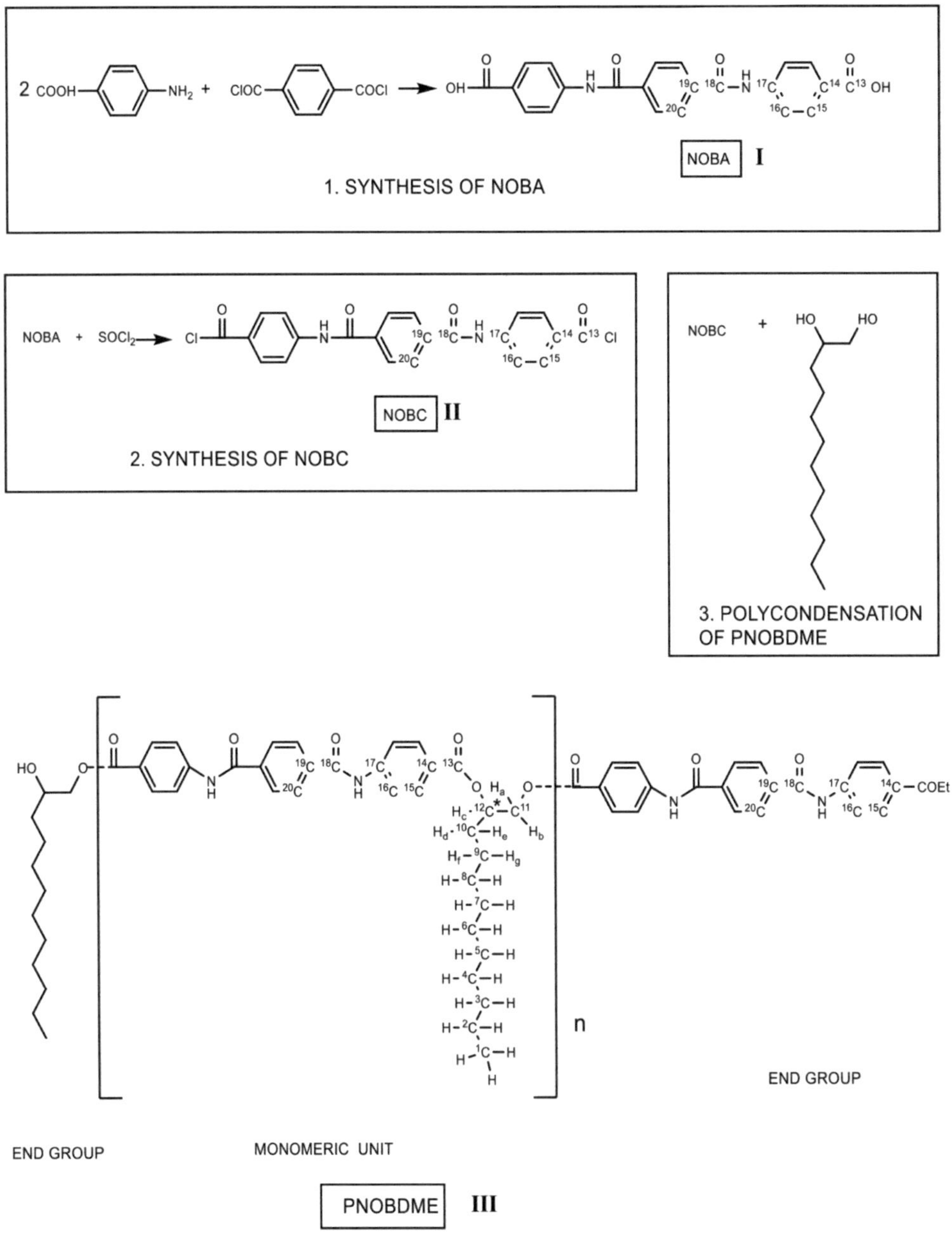

Figure 8.
Scheme of the synthetic way to attain PNOBDME.

The new cationic cholesteric LC polymers so designed were synthesized as follows: PTOBDME-choline [$(C_{34}H_{36}O_8)_n$–$C_5H_{13}N$]; PTOBEE-choline [$(C_{26}H_{20}O_8)_n$ –$C_5H_{13}N$]; PTOBDME-ammonium [$(C_{34}H_{36}O_8)_n$–$C_5H_{13}N$]; PTOBEE-ammonium [$(C_{26}H_{20}O_8)_n$–$C_5H_{13}N$]; PTOBUME-amide [$(C_{33}H_{33}O_9N)_n$]; and PTOBEE-amide $(C_{26}H_{19}O_9N)_n$ [29].

These cationic polymers proved to be biocompatible, able to form polyplexes and capable of successfully condensing and transfecting the DNA into the nucleus cell, protecting DNA from inactivation by blood components. The complexes are sensitive to pH changes, possessing substantial buffering capacity below physiological pH. Their efficiency relies on extensive endosome swelling and rupture that provides an escape mechanism for the polycation/DNA complexes [30].

Since we are mainly interested in the design and chemical modifications of multifunctional cholesteric LC polyesters involving new properties but holding the precursor helical macromolecular structure, new cationic functionalization is held by introducing amide groups *para*-substituting the two ester groups in the central benzene ring of the terephthalate unit, along the main chain [31].

The final formulations of the new monomers, called PNOBDME and PNOBEE, are shown in **Figure 7(a)** and **(b)**, respectively. The hydrogen and carbon atoms have been numbered as precursors PTOBDME [16, 17] and PTOBEE [11, 15], respectively.

The synthetic way is based on our previous experience for the attainment of cholesteric LC polyesters and the condensation reaction reported by Sek et al. to obtain polyesteramides [32]. The intermediate acid chloride yield obtained here is lower than the precursor polyesters obtained. The synthetic way of polyesteramides PNOBDME $[C_{34}H_{38}N_2O_6]_n$ and PNOBEE $(C_{26}H_{22}N_2O_6)_n$ is given in **Figures 8** and **9**.

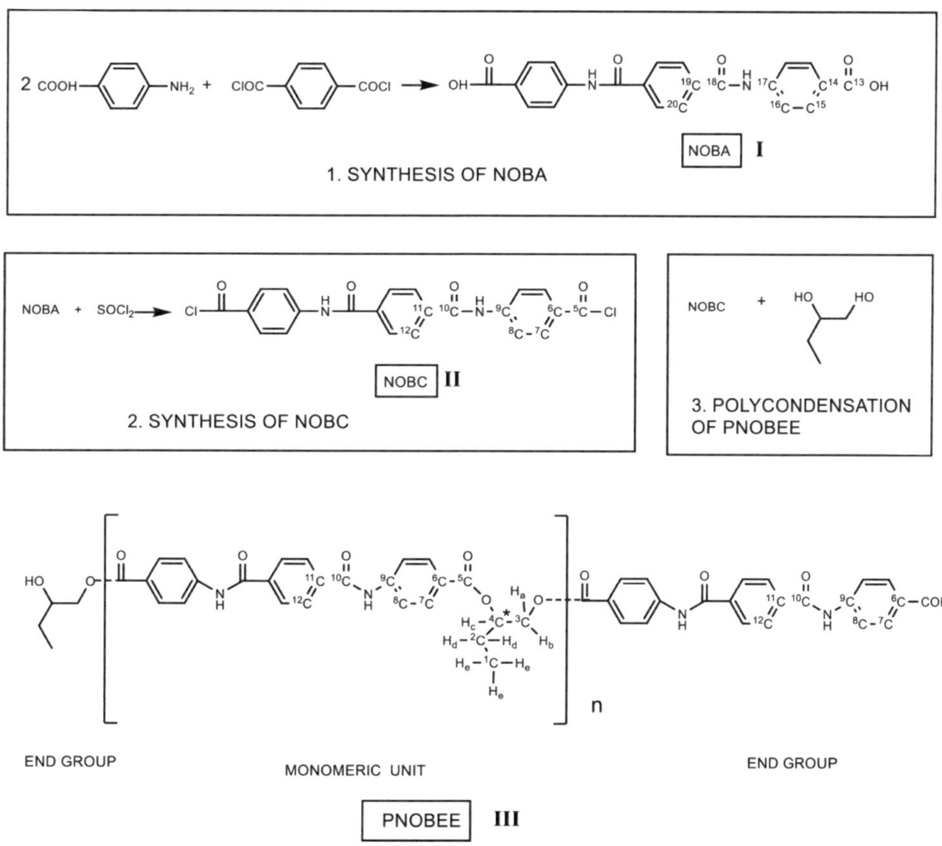

Figure 9.
Scheme of the synthesis of PNOBEE.

The structure of the polymers so obtained could be confirmed by ^1H, ^{13}C, COSY and HSQC NMR [31]. The NMR shifts were assigned according to our previous notation. Their thermal stability is studied by thermogravimetric (TG) and differential scanning calorimetry (DSC) analysis.

2. Experimental

2.1 Materials

2.1.1 Synthesis of PNOBDME {Poly[oxy(1,2-dodecane)-oxy-carbonyl-1,4-phenylene-amine-carbonyl-1,4-phenylene-carbonyl-amine-1,4-phenylene-carbonyl]}, $(C_{34}H_{38}N_2O_6)_n$

PNOBDME (III in **Figure 8**) was obtained through condensation reaction between 4, 4′-(terephthaloyl-diaminedibenzoic chloride) (NOBC) (II in **Figure 8**) and the racemic mixture of DL-1,2-dodecanediol. Similar notation to precursor cholesteric liquid crystal PTOBDME [16, 17] obtained by a similar method has been used.

2.1.1.1 Preparation of NOBA

Solutions of 0.1 mol terephthaloyl chloride in 200 ml carbon tetrachloride and 0.2 mol NaOH in water were added while stirring at room temperature for 15 min to a solution of 0.22 mol of 4-aminobenzoic acid and 0.2 mol NaOH in 400 ml water (Milli-Q grade) (4–4′-(terephthaloyl-diaminedibenzoic acid)) (NOBA), (I in **Figure 8**). Stirring was continued for 12 hours. Sediment was separated, filtered, washed several times with 40 ml of cold water, dried, comminuted and transferred to a vessel where it was mixed for 3 hours with 300 ml of hydrochloric acid. The product was filtered, washed several times with 40 ml of cold water, dried and comminuted. Yield is 28 g (70%).

2.1.1.2 Preparation of NOBC

In the course of 25 min, 15 g NOBA was added to 350 ml thionyl chloride, while stirring rapidly at room temperature (4- 4′-(terephthaloyl-diaminedibenzoic chloride)) (NOBC) (II in **Figure 8**). The solution was boiled with the reflux condenser. When the release of HCl ended and most of the sediment had dissolved, the hot solution was filtered and cooled down to 0°C for a day. The obtained product was separated, filtered, vacuum dried and recrystallized in chloroform. Yield is 7.2 g (48%).

2.1.1.3 Preparation of PNOBDME

A mixture of 0.017 mol NOBC and 0.017 mol DL-1,2-dodecanediol was added to 44 ml of diphenyl oxide. Purge with dry nitrogen was used for 25 min at room temperature, and then, while maintaining the gas current, the flask was transferred to a bath containing a high-temperature heat-transfer agent. The polycondensation was carried out for 3 hours and 30 min at 200°C. The reaction finished when the liberation of HCl ended. The result of the polycondensation reaction was poured into 500 ml of toluene, decanting PNOBDME, which was filtered, washed three times with ethanol and vacuum dried. The second fraction of PNOBDME precipitated of the filtrated toluene after 22 weeks was also filtered, washed with ethanol and vacuum dried. Yield first fraction is 2.6 g (25.5%); yield first and second fraction is 3.1 g (30.4%).

2.1.2 Synthesis of PNOBEE {Poly[oxy(1,2-butylene)-oxy-carbonyl-1,4-phenylene-amine-carbonyl-1,4-phenylene-carbonyl-amine-1,4-phenylene-carbonyl]}: $(C_{26}H_{22}N_2O_6)_n$

PNOBEE (III in **Figure 9**) was obtained through condensation reaction between (4- 4′-(terephthaloyl-diaminedibenzoic chloride)) (NOBC) (II in **Figure 9**) and the racemic mixture of DL-1,2-butanediol. Notation of cholesteric liquid crystal precursor PTOBEE has been used.

NOBC was synthesized by the reaction between NOBA, (4–4′-(terephthaloyl-diaminedibenzoic acid)) (NOBA) (I in **Figure 9**), and $SOCl_2$ and recrystallized in chloroform; previously NOBA was obtained by interface condensation between terephthaloyl chloride and 4-aminobenzoic acid.

2.1.2.1 Preparation of PNOBEE

A mixture of 0.015 mol NOBC and 0.015 mol DL-1,2-butanediol was added to 39 ml of chloronaphthalene. Purge with dry nitrogen was used for 25 min at room temperature, and then, while maintaining the gas current, the flask was transferred to a bath containing a high-temperature heat-transfer agent. The polycondensation was carried out for 180 min at 200°C. The reaction finished when the liberation of HCl ended. The result of the polycondensation reaction was poured into 500 ml of toluene, decanting PNOBEE, which was filtered, washed with ethanol and vacuum dried. The second fraction of PNOBEE precipitated of the filtrated toluene after 22 weeks which was also filtered, washed with ethanol and vacuum dried. Yield is 2.9 g (46.5%).

The structures of NOBA and NOBC were confirmed by ^{1}H-RMN, ^{13}C-NMR, COSY, TOCSY, NOESY and HSQC registered in DMSO-d_6 at 25°C in a Bruker 300 MHz NMR spectrometer. The structure of PNOBDME was studied by ^{1}H-RMN, ^{13}C-NMR, COSY and HSQC, obtained in VARIAN 400 MHz and 500 MHz spectrometers, also at room temperature [31].

2.1.2.2 Starting materials

Terephthaloyl chloride from Sigma-Aldrich Chemie GmBH (Steinheim, Germany); carbon tetrachloride from Panreac Química (Montcada i Rexach, Barcelona, Spain); NaOH from Panreac Química (Montcada i Rexach, Barcelona, Spain); 4-aminobenzoic acid from Sigma-Aldrich Chemie GmBH (Steinheim, Germany); hydrochloric acid from Normapur VWR International (Fontenay-sous-Bois, France); thionyl chloride from Sigma-Aldrich Chemie GmBH (Steinheim, Germany); chloroform from SDS Votre Partenaire Chimie (Peypin, France); DL-1,2-dodecanediol from Fluka Chemie GmBH (Buchs, Switzerland); diphenyl oxide from Sigma-Aldrich Chemie GmBH (Steinheim, Germany); nitrogen from Praxair (Madrid, Spain); toluene from Merck KGaA (Darmstadt, Germany).

DL-1,2-butanediol from Fluka Chemie GmBH (Buchs, Switzerland); chloronaphthalene from Sigma-Aldrich Chemie GmBH (Steinheim, Germany).

The solvent used in NMR for all cases was DMSO-d_6 from Merck KGaA (Darmstadt, Germany) and DMSO for optical rotatory dispersion (ORD), from Scharlau Chemie.

3. Characterization techniques

Thermal stability was studied on a Mettler TA4000-TG50 at a heating rate of 10°C/min with nitrogen purge between 30 and 600°C. Thermal behaviour was analyzed by DSC in a Mettler TA4000/DSC30/TC11 calorimeter, with a series of heating/cooling cycles in a temperature range between 0 and 230°C.

Microcalorimetry was estimated in a MicroCal Inc., model MCS-DSC, within a range of temperature 4–120°C, at a heating rate of 10–20°C/h, and a volume of sample 1.5 ml.

Optical rotatory dispersion was measured in a Perkin Elmer 241 MC polarimeter, at 25°C in DMSO. Conditions used: λ_{Na} = 589 nm, slit = 5 mm, integration time = 50 s; λ_{Hg} = 574 nm, slit = 14 mm, integration time = 50 s; λ_{Hg} = 546 nm, slit = 30 mm, integration time = 50 s; λ_{Hg} = 435 nm, slit = 5 mm, integration time = 50 s; and λ_{Hg} = 365 nm, slit = 2.5 mm, integration time = 50 s.

Morphology was evaluated in an environmental scanning electron microscope (ESEM), PHILIPS XL30.

Simultaneous SAXS/WAXS of PNOBDME were performed at 16.1.1 beamline of the synchrotron radiation source (SRS) at Daresbury Laboratory, Warrington, UK, with a monochromatized beam (λ = 1.4 Å). Both WAXS and SAXS detectors were lineal. HDPE was used to calibrate the WAXS data and wet collagen (rat tail tendon, d = 676.08 Å) to calibrate the *q*-axis of the SAXS detector ($q = 4\pi \sin \theta/\lambda$), where the scattering angle is defined by 2θ. The experimental data were corrected for background scattering, sample absorption and positional lack of linearity of the detector, with the help of ATSAS [23, 24]. The samples dispersed in dichloromethane solution were applied dropwise on the sample holder and the solvent let to evaporate.

3.1 Thermal behaviour of PNOBDME and PNOBEE

Figure 10(a) shows the thermogravimetric curve of polyesteramide PNOBDME first fraction. A 5 and 10% weight loss is observed, respectively, at 282 and 310.3°C, increasing the thermal stability range of precursor polyester PTOBDME, with 10% weight loss percentage at 280°C [16].

In **Figure 10(b)**, the PNOBEE thermal stability can be observed. A 10% weight loss is registered at 330°C, due to thermal decomposition, a value higher than those observed for PTOBEE (280°C) [11] and PNOBDME (310.3°C). The entrance of the amide group in the mesogen causes an increase of thermal stability with respect to PTOBEE.

In **Figure 11(a)**, the DSC analysis of PNOBDME first fraction is exhibited and its Microcalorimetry curve appears in **Figure 11(b)**.

During the first heating run of the DSC of PNOBDME, performed at 10°C/min rate, **Figure 11(a)**, a glass transition around 62.5°C is observed, together with a

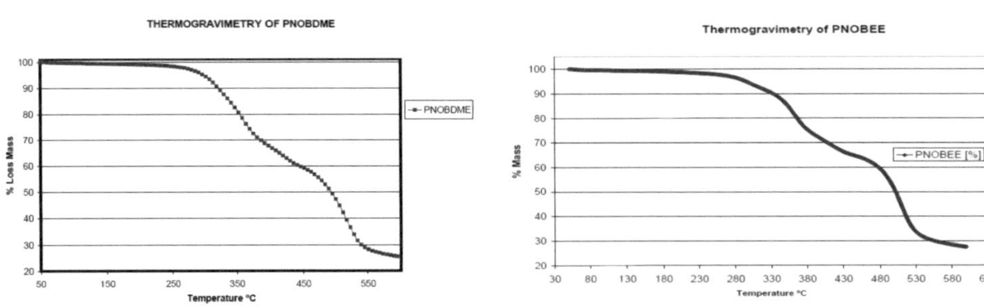

(a) (b)

Figure 10.
(a) Thermogravimetric curve of PNOBDME first fraction; (b) thermogravimetric curve of PNOBEE first fraction.

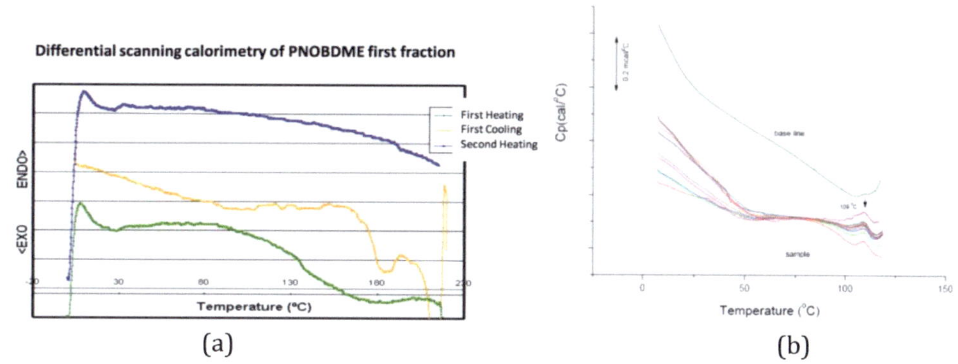

Figure 11.
(a) DSC analysis of PNOBDME first fraction; (b) microcalorimetry of PNOBDME.

small broad endothermic peak centred at 156°C. During the cooling, an exothermic peak at 183°C is indicative of crystallization from the mesophase, a higher value than that of PTOBDME (149°C). In the second heating, a glass transition is observed around 71.6°C and two broad and small endothermic peaks at 108.7 and 188.3°C.

Subsequently, PNOBDME was heated up to 230°C, at 10°C/min, cooled to 190°C and isothermally heated for 2 hours, then cooled to 30°C at 10°C/min and finally heated again to 230°C, at 10°C/min. Although the isothermal treatment at 190°C, after cooling from 230°C, should have produced an induced crystallization process (endothermic peak due to the polymer transition to mesophase), only a small endothermic peak at 109°C is observed not caused by the isothermal cooling. The endothermic transition from crystal to mesophase was also confirmed at 109°C by microcalorimetry, **Figure 11(b)**.

The DSC curve of PNOBEE is shown in **Figure 12**. During the first heating run (a), at a 10°C/min rate, a glass transition around 55°C is observed. A very broad endothermic peak centred at 185.3°C is interpreted as a fusion associated with a transition from crystal to liquid crystal. Another endothermic peak at 233.7°C is observed near the beginning of thermal decomposition. In the cooling run (b), two small exothermic peaks observed at 205.6 and 183.0°C are interpreted as crystallization processes from the mesophase. In the second heating (c), no transition is observed.

Compared to PTOBEE, with a transition to mesophase at 150°C and with exothermic crystal formation at 110°C during isothermal heating, a remarkable difference is observed by the substitution of ester groups by amide in the mesogen, increasing the stability range.

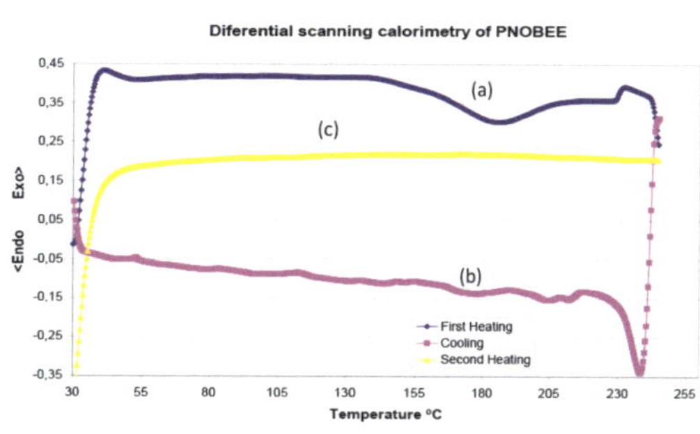

Figure 12.
DSC analysis of PNOBEE first fraction. (a) First heating run of the original sample; (b) subsequent cooling down; (c) second heating run.

3.2 Morphology of PNOBDME

The morphology of powdered PNOBDME without any previous treatment has been studied by SEM. In **Figure 13**, six details of the powder sample (a to f) are shown of the homogeneous spherical clusters of about 5 μm in diameter homogeneously dispersed.

3.3 Optical activity

Although synthesized from starting racemic materials, PNOBDME showed unexpected chirality. The first fraction of the polymer did not show a net optical activity but values fluctuated from positive to negative, but the second fraction presented a low but constant value +1.02°, at 598 nm; +1.65°, at 579 nm; and +2.9°, at 435 nm and a very high optical activity value between +600° and +950°, at 365 nm, depending on temperature. The same behaviour had been also observed in PTOBDME and PTOBEE.

The variation of the optical rotatory dispersion values (α) by the effect of time of the second fraction of PNOBDME is expressed in **Figure 14** as molar optical rotation $[\Phi] = [\alpha]\ M/100$, at 365 nm, at two different temperatures, 35°C and 25°C, being *M* the molecular weight of the polymer repeating unit.

At 35°C, the ORD of PNOBDME increases with time to a value approximate of 600°, preserved between 90 and 180 min. After that, it increases again to reach a value of 950°, getting stabilized after 360 min. After 120 min at 25°C, the ORD reaches a value of 600° conserved up to 9 hours.

In both cases, once the ORD value was stabilized to 600° and 950°, if the wavelength of the lamp changed from 365 to 435 nm and quickly returned to 365 nm, the ORD value initially decreased to +8.6 but recovered its value. This phenomenon is being reversible. The variation of ORD with time has been described in helical polyguanidines synthesized either from chiral monomers or from achiral monomers with chiral catalysts [33–39]. At the end of this article, the optical activity of PNOBEE has not been studied.

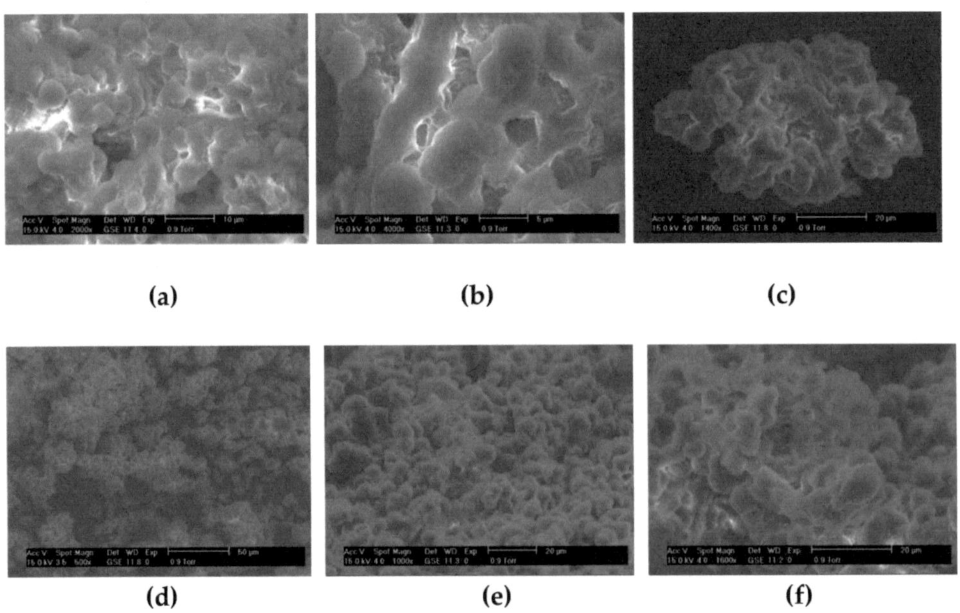

Figure 13.
SEM images of powdered PNOBDME.

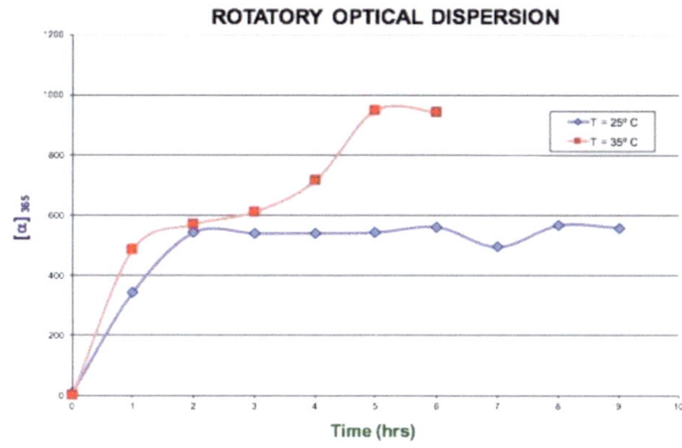

Figure 14.
Optical rotatory dispersion of PNOBDME second fraction at 25C and 35°C.

3.4 Molecular mechanics simulation of PNOBDME.

The structural fragment including a chiral secondary alcohol and a primary alcohol group (a beta-chiral 1,2-diol) is particularly interesting since it is present in many relevant natural products, such as sugars, nucleosides, glycerides [40], chiral nanostructures from helical polymers and metallic salts [41].

In the case of PNOBDME, the fragment in the spacer including the secondary alcohol group, bonded to chiral $^{12}C^*$, and the primary alcohol, bonded to prochiral ^{11}C, is shown in **Figure 15** for the R enantiomer of $^{12}C^*$.

Molecular mechanics always predict helical macromolecular structures along the main chain for PNOBDME and PNOBEE, as formulated in **Figure 7**. Instead, no helical polymer models were attained in the computational calculations when the amide group enters along the lateral side chains.

Molecular mechanics modeling was performed for the PNOBDME monomer with Materials Studio Windows v. 2019 [42]. COMPASS-II force field was loaded, including both atomic mass and charge. A model of the monomer is shown in **Figure 16(a)**, with optimized geometry to a minimum of energy, -95 Kcal/mol. Monomer polymerization was simulated by defining the ^{11}C atom as the *head atom*, within the *repeating unit*, and the O atom bonded to $^{13'}C$, as the *tail atom*, being $^{12}C^*$ the chiral centre. Homopolymerization was then simulated with head-to-tail orientation and torsion angle between monomers fixed to 180°. Isotacticity was finally imposed on the polymer chain. The helical polymer model so obtained along the main chain is shown in **Figure 16(b)**. The perpendicular cross-section appears in **Figure 16(c)**.

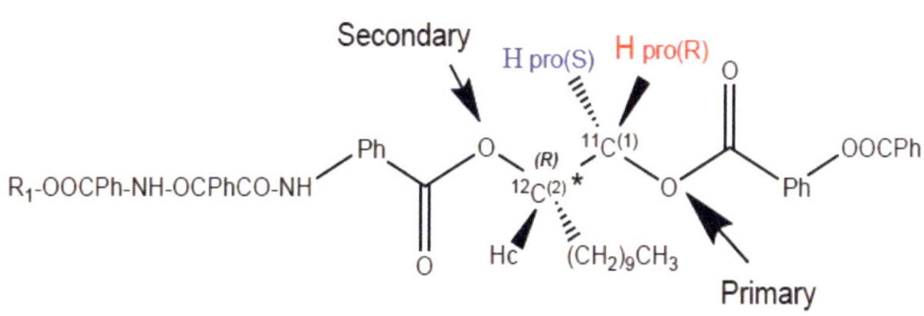

Figure 15.
Scheme of the spacer of PNOBDME including the two alcohol groups.

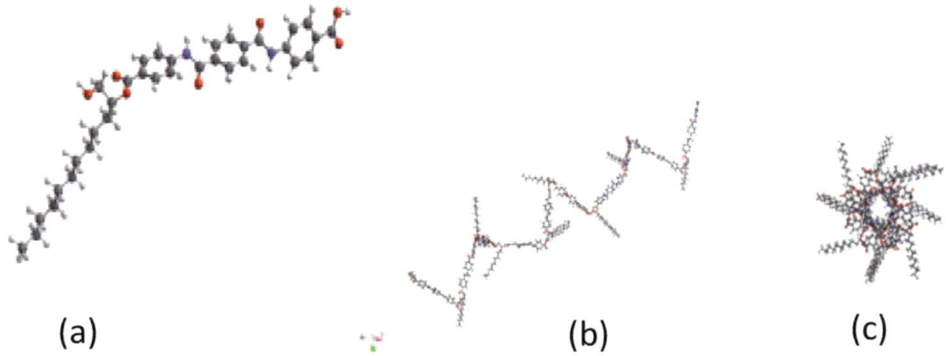

Figure 16.
Molecular simulation of PNOBDME monomer: (a) minimum energy MM model; (b) isotactic $[PNOBDME]_{10}$; (c) cross-sectional view.

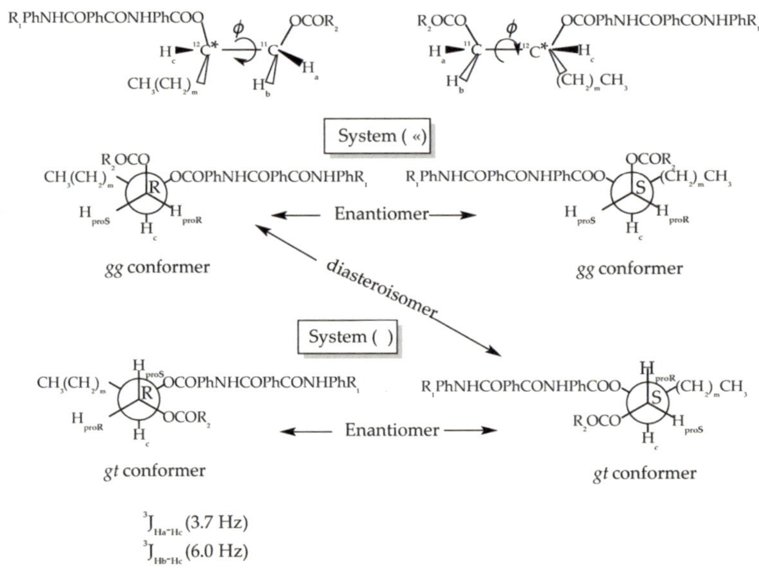

Figure 17.
The relationship between the four helical diastereomers gg and gt of the R and S enantiomers of PNOBDME and PNOBEE through the ^{11}C–$^{12}C^*$ bond, and ^{3}C–$^{4}C^*$ (torsion φ), respectively.

3.5 Conformational analysis of PNOBDME and PNOBEE

The tetrahedral carbon atoms **11C** in PNOBDME, allocated in *α* with respect to the asymmetric carbon atom $^{12}C^*$ (**Figure 7(a)** with m = 9) and **3C** in PNOBEE, allocated in *α* with respect to chiral $^{4}C^*$ (**Figure 7(b)** with m = 1), along the polymer backbones, are referred as *prochirals*, since both could be converted into a chiral centre by arbitrarily changing only one attached H group to a deuterium atom (D with higher priority than H). Depending on the configuration, R/S, of the so created chiral centre, the H atom ideally deuterated, would be labeled as *pro-R/S*.

The two hydrogen atoms on the *prochiral ^{11}C carbon atom*, H_a and H_b, in PNOBDME, can be described as *prochiral hydrogens*. Prochiral hydrogens can be also designated as *diastereotopic*. Their indistinguishable signals by ^1H-NMR split then into two signals easily differentiated. The same effect is observed for H_d and H_e, bonded to prochiral ^{10}C, and for H_f and H_g, bonded to prochiral ^{9}C.

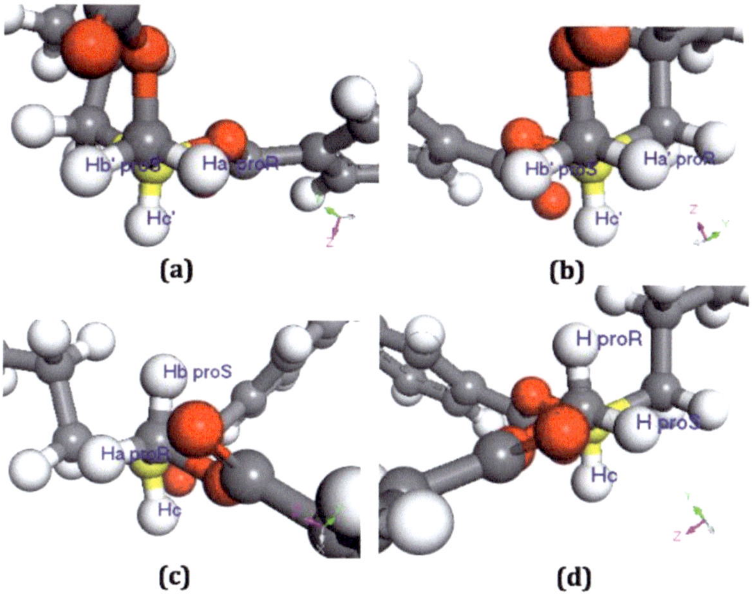

Figure 18.
Molecular model details of a PNOBDME dimer. View along $^{11}C{-}^{12}C^*$ *bond (perpendicular to the paper), with (R) and (S) absolute configuration of* $^{12}C^*$ *(in yellow behind* ^{11}C*) for (a) Rgg-diasteroisomer; (b) Sgg-diasteroisomer; (c) Rgt-diasteroisomer; (d) Sgt diasteroisomer.*

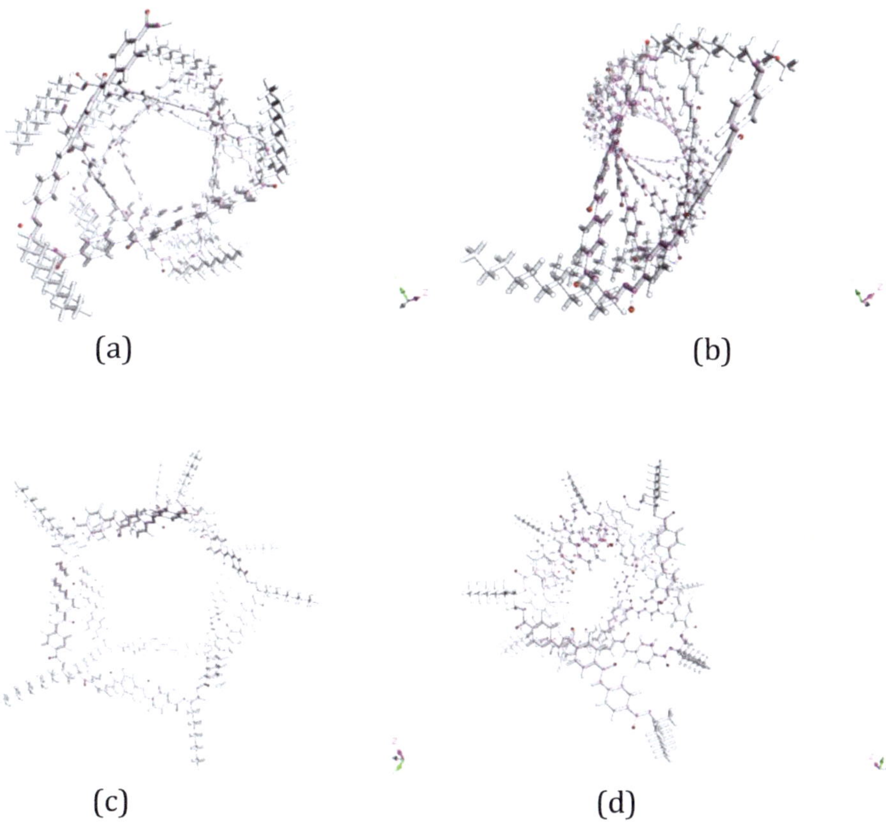

Figure 19.
Molecular model details of PNOBDME polymers: (a) Poly10_PNOBDME_Rgg; (b) Poly10_PNOBDME_Sgg; (c) Poly10_PNOBDME_Rgt; (d) Poly10_PNOBDME_Sgt.

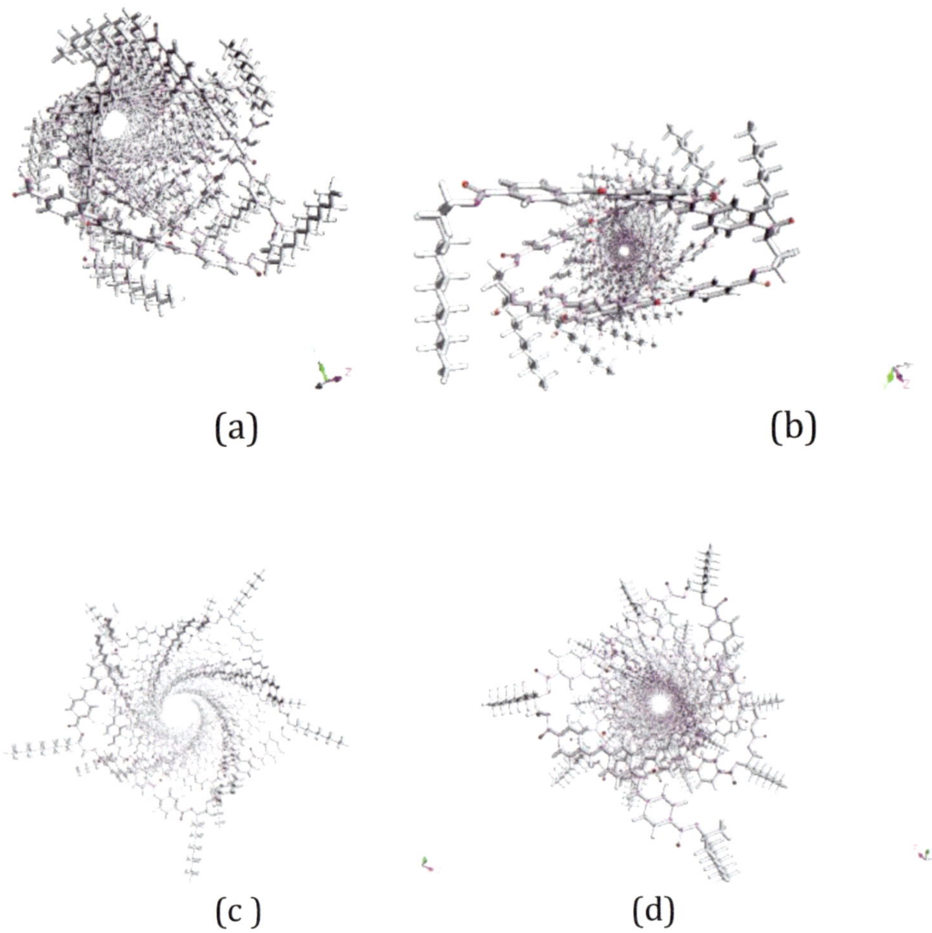

Figure 20.
Molecular model details of PNOBDME polymers: (a) Poly60_PNOBDME_Rgg; Poly60_PNOBDME_Sgg; (c) Poly60_PNOBDME_Rgt; (d) Poly60_PNOBDME_Sgt.

Equally the two hydrogen atoms on *prochiral ^3C carbon atom*, H_a and H_b, of PNOBEE, are considered as *prochiral hydrogens.*

Two independent sets of signals are experimentally observed by ^{1}H-NMR for each enantiomer of PNOBDME and PNOBEE [31]. They are related to the two possible staggered diastereomeric conformers, *gg* and *gt* of torsion φ along the ^{11}C–^{12}C* bond in PNOBDME and ^{3}C–^{4}C* in PNOBEE, along the polymer backbone. One of these two systems is designated with an apostrophe (') and the other is designed without an apostrophe ().

The combination of a helix with two screw senses and the two absolute configurations by the presence of the asymmetric carbon atom provides four diastereomeric structures. There are two pairs of enantiomers each with two independent sets of ^{1}H-NMR signals. The four diastereomers of PNOBDME are depicted in **Figure 17**.

The existence of two independent conformers had also been observed for each enantiomer of PTOBDME and PTOBEE. It was also related to the presence of helical structures, the Cotton effect and the sign of the helicity in the case of 1-2 di-O-benzoylated sn-glycerols [33–39].

Details of molecular models for *gg* and *gt* conformers of a dimer of PNOBDME are shown (**Figure 18**), projected along the ^{11}C–^{12}C* bond and torsion φ

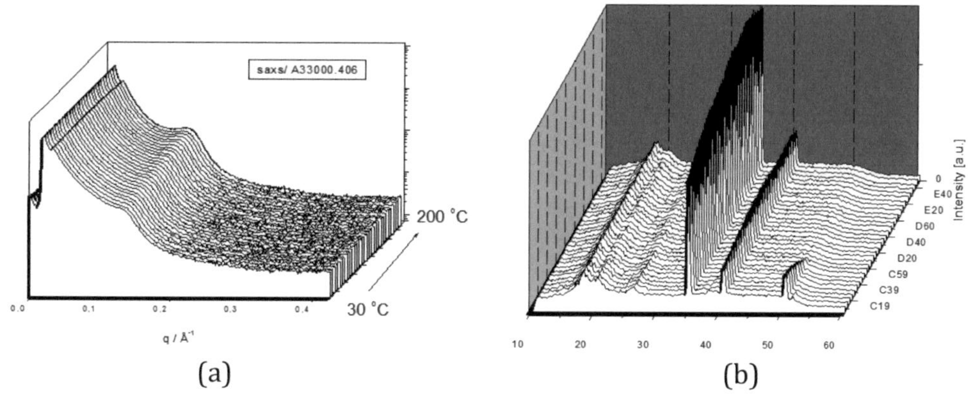

Figure 21.
Simultaneous SAXS (a)/WAXS (b) patterns of PNOBDME.

(perpendicular to the paper) with $^{12}C^*$ (bonded to H_c) having *R* and *S* absolute configuration, in yellow, behind ^{11}C (bonded to H_a and H_b).

Helical polyesteramide $(PNOBDME)_{10}$ molecular models obtained with the four diastereomers, after minimizing the corresponding monomers are described in **Figure 19**.

Polymer models with 60 monomers of the 4 are exhibited in **Figure 20**.

3.6 Simultaneous SAXS/WAXS of PNOBDME

Figure 21(a) and **(b)** shows the simultaneous SAXS/WAXS patterns of PNOBDME registered during heating from 30–200°C, with a synchrotron radiation source.

The SAXS spectra show two sharp order reflections at *q* value of 0.018 $Å^{-1}$ (55.5 Å) and 0.029 $Å^{-1}$ (34.48 Å).

Four WAXS peaks, at $2\theta \cong 18.39^{\circ}$ (d = 4.38 Å); 18.83° (d = 4.28 Å); 35.17 (d = 2.31 Å); and 40.75 (d = 2.01 Å) always remaining present in the entire temperature range, were assigned to the cholesteric mesophase.

Peaks at $2\theta \cong 26.39$ (d = 3.07 Å) and 50.71 (d = 1.73 Å) disappearing at about 90°C during the heating range are attributed to crystal 3D phase.

4. Conclusions

The synthetic multifunctional cholesteric liquid crystal polyesteramides designed as PNOBDME $(C_{34}H_{38}N_2O_6)_n$ and PNOBEE $(C_{26}H_{22}N_2O_6)_n$ are reported as chemical modifications of multifunctional cholesteric LC polyesters, involving new properties but holding the precursor helical macromolecular structure.

Molecular mechanics models of the new polymers show helical polymeric rigid chains. Homopolymerization was simulated with head-to-tail orientation and torsion angle between monomers fixed to 180°. Isotacticity was finally imposed on the polymer chains, explained in terms of the higher reactivity of the primary hydroxyl regarding the secondary one in the glycol through the polycondensation reaction.

Two independent sets of signals experimentally observed by ^1H-NMR for each enantiomer of PNOBDME and PNOBEE (while the R/S ratio of asymmetric carbon atoms remained 50:50) are related with two possible staggered diastereomeric conformers, *gg* and *gt* of torsion φ, containing the asymmetric carbon atom in the spacer, along the polymer backbone. One of these two systems is designated with an apostrophe ('), and the other is designed without an apostrophe (). The combination of a helix with two screw senses and the two absolute configurations by the presence of the asymmetric carbon atom provides four diastereomeric structures.

Molecular models of the four diastereomers *Rgg*, *Sgg*, *Rgt* and *Sgt* are provided as well as their polymer chains with 10 and 60 monomeric units.

The thermal behaviour of the new synthesized cholesteric liquid crystal polyesteramides has been studied by TG and DSC. An endothermic peak assigned to the first-order transition from crystal to liquid crystal mesophase is observed in both polymers.

ORD values are provided for polyesteramide PNOBDME. The first fraction of the polymer did not show a net optical activity but values fluctuating from positive to negative, but the second fraction presented low positive values +1.02°, at 598 nm; +1.65°, at 579 nm; and +2.9°, at 435 nm but very high optical activity +600° to +950°, at 579 nm, when increasing temperature from 25–35°C.

Morphology of powdered PNOBDME by ESEM shows spherical clusters of about 5 μm in diameter, homogeneously dispersed.

Simultaneous SAXS/WAXS patterns of PNOBDME are registered during heating from 30 to 200°C, with a synchrotron radiation source.

Acknowledgements

We thank CSIC for their facilities. We acknowledge support of the publication fee by the CSIC Open Access Publication Support Initiative through its Unit of Information Resources for Research (URICI).

Author details

Mercedes Pérez Méndez[1*] and José Fayos Alcañiz[2]

1 Instituto de Ciencia y Tecnologia de Polimeros (CSIC), Madrid, Spain

2 Instituto de Química Física Rocasolano (CSIC), Madrid, Spain

*Address all correspondence to: perezmendez@ictp.csic.es

References

[1] Schrödinger E. What Is Life? The Physical Aspect of the Living Cell. New York: Cambridge University Press; 1945

[2] Macías Barber E. El cristal aperiódico de la vida. Revista de la Unión Iberoamericana de la Sociedad de Física. 2006;**2**:7-16

[3] Franklin RE, Goslin RG. Molecular configuration in sodium thymonucleate. Nature. 1953;**171**:740-741

[4] Watson JD, Crick FHC. Molecular structure of nucleic acids: A structure for deoxyribose nucleic acid. Nature. 1953;**171**:737-738

[5] Wilkins MHF, Stokes AR, Wilson HR. Molecular structure of deoxypentose nucleic acids. Nature. 1953;**171**:738-740

[6] Ringsdorf H, Schlarb B, Venzmer J. Molecular architecture and function of polymeric oriented systems: Models for the study of organization, surface recognition, and dynamics of biomembranes. Angewandte Chemie (International Ed. in English). 1988;**27**:113

[7] International Union of Crystallography. Report of the Executive Committee for 1991. Acta Crystallographica. 1992;**A48**:922

[8] Zanchetta G. Liquid crystalline phases in oligonucleotide solutions [thesis]. University of Milan; 2007

[9] Kornyshev AA, Leikin S, Malinin SV. Chiral electrostatic interaction and cholesteric liquid crystals of DNA. European Physical Journal E: Soft Matter and Biological Physics. 2002;**7**:83-93

[10] Lai SL, Hartono D, Yang K-L. Self-assembly of cholesterol DNA at liquid crystal/aqueous interface and its application for DNA detection. Applied Physics Letters. 2009;**95**:153702

[11] Pérez-Méndez M, Marco C. New synthesis, thermal properties and texture of cholesteric poly[ethyl ethylene 4,4′-(terephthaloyldioxy)dibenzoate]. Acta Polymerica. 1997;**48**:502-506

[12] Pérez-Méndez M, Marco Rocha C. Preparing cholesteric liquid-crystals - by adding acid dichloride and butanediol to chloro-naphthalene, heating in nitrogen, decanting into toluene, etc. Patent family: Patent with no. EP1004650-A; WO9831771-A; WO9831771-A1; AU9854863-A; ES2125818-A1; ES2125818-B1; EP1004650-A1; US6165382-A; MX9906732-A1; JP2001513827-W; AU739076-B; EP1004650-B1; DE69824182-E

[13] Bilibin AY, Ten'kovtsev AV, Piraner ON, Skorokhodov SS. Synthesis of high molecular weight liquid crystal polyesters based on a polycondensation mesogenic monomer. Polymer Science U.S.S.R. 1984;**26**(12):2882-2890

[14] Bilibin AY, Skorokhodov SS. Rational path of the synthesis of liquid-crystalline high molecular weight polyesters and their properties in solution. Macromolecular Symposia. 1989;**29**:9-23

[15] Perez-Mendez M, Marsal R, Garrido L, Martin-Pastor M. Self-association and stereoselectivity in a chiral liquid-crystal cholesteric polymer formed under achiral conditions. Macromolecules. 2003;**36**:8049-8055

[16] Fayos J, Sánchez-Cortés S, Marco C, Pérez-Méndez M. Conformational analysis and molecular modeling of cholesteric liquid-crystal polyesters based on XRD, Raman and transition thermal analysis. Journal of Macromolecular Science, Physics. 2001;**B40**(3&4):553-576

[17] Pérez Méndez M, Sanguino J. Cholesteric liquid-crystal copolyester,

poly[oxycarbonyl-1,4-phenylene-oxy –1,4 terephthaloyl- oxy-1,4-phenylene- carbonyloxy (1,2-dodecane)] [C34H36O8]n, synthesized from racemic materials: Kinetics, structure and optical characterization. Journal of Engineering Research and Applications. 2015;**5**(7, (Part-2)):48-62

[18] Boiko N, Shibaev V. Cholesteric polymer liquid crystals and their optical properties. International Journal of Polymeric Materials. 2000;**45**(3-4):533-583

[19] Sánchez-CortésS,Marsal-BerenguelR, Pérez-Méndez M. Adsorption of a cholesteric liquid-crystal polyester on silver and gold nanoparticles and films studied by surface-enhanced Raman scattering. Applied Spectroscopy. 2004;**58**(5):562-569

[20] Pérez-Méndez M, Fayos J, Mateo CR. Chapter 24: Self-assembly of cholesteric liquid-crystal polyesters and their stereoselective interaction with liposomes of DMPC. In: Pályi G, Zucchi C, Caglioti L, editors. Advances in Biochirality. Oxford, UK: Elsevier Science Ltd.; 1999

[21] Pérez-Méndez M, Areso S, Alarcón Vaquero A, Elorza B, Malfois M. Effect of polymer addition to lipid membranes as potential drug delivery systems. Structure and dynamics of their interaction, Annual Report EMBL; 1998. p. 372

[22] Perez-Mendez M, Fayos J, Blanch GP, Sánchez Cortés S. Biofunctionalization of cholesteric liquid-crystal helical polymers. Nanocarriers. In: Nalwa HS, editor. Encyclopedia of Nanoscience and Nanotechnology. ACS. Valencia, California, USA: American Scientific Publishers. Vol. 11. 2011. pp. 547-580. ISBN: 1-58883-160-4

[23] Konarev PV, Petoukhov MV, Volkov VV, Svergun DI. ATSAS 2.1, a program package for small-angle scattering data analysis. Journal of Applied Crystallography. 2006;**39**(Part 2):277-286. DOI: 10.1107/S0021889806004699

[24] Kostorz G, Svergun DI, Koch MHJ, Timmins PA, May RP, editors. Small angle X-ray and neutron scattering from solutions of biological macromolecules. In: IUCr Texts on Crystallography; No. 19. Oxford University Press; 2013. p. 358. ISBN: 978-0-19-963953-3. Published to Oxford Scholarship Online: December 2013: DOI: 10.1093/acprof:oso/9780199639533.001.0001

[25] Anderson WF. Human gene therapy. Nature. 1998;**392**(6679 Suppl):25-30

[26] Ibraheem D, Elaissari A, Fessi H. Gene therapy and DNA delivery systems. International Journal of Pharmaceutics. 2014;**459**(1-2):70-83

[27] Pérez Méndez M, Marsal Berenguel R, Sánchez Cortés S. New non-viral vectors based on biocompatible liquid-crystal polymers for the carriage and delivery of biomacromolecules and insoluble drugs as a strategy. Revista de Oncología. 2002;**4**(Supl. 1):177

[28] Pérez Méndez M, Marsal Berenguel R, Funari SS. Structural characterization of the interaction between cholesteric liquid-crystal polymers and molecules of biological interest. HASYLAB Annual Report. 2003:11150. Available from: http://hasyweb.desy.de/science/annual_reports/2003_report/part1/contrib/47/11150.pdf

[29] Pérez Méndez M. Chapter 2: Synthetic cationic cholesteric liquid crystal polymers, liquid crystals. In: Choudhury PK, editor. Recent Advancements in Fundamental and Device Technologies. Rijeka: InTech Open; 2018. pp. 7-33. Available from: https://cdn.intechopen.com/pdfs-wm/58375.pdf

[30] Pérez Méndez M, Rodríguez Martínez D, Fayos J. Structure of non-viral vectors based on cholesteric liquid-crystal polymers by SAXS. International Journal of Advancement In Engineering Technology, Management and Applied Science (IJAETMAS). 2016;**03**(11):27-41. ISSN: 2349-3224

[31] Pérez Méndez M. Biocompatible, nanostructured, chiral polyesteramides: PNOBDME (C34H38N2O6)n and PNOBEE (C26H22N2O6)n synthesized and characterised as cholesteric liquid crystals. International Journal of Engineering Research and Applications (IJERA). 2019;**9**(6 (Part –1)):52-66. DOI: 10.9790/9622- 0906015267

[32] Sek D, Wolinska A, Janeczek H. Structure-liquid crystalline properties relationship of polyesteramides. Journal of Polymer Materials. 1986;**3**:225-233

[33] Uzawa H, Nishida Y, Ohrul H, Meguro H. Application of the dibenzoate chirality method to determine the absolute configuration of glycerols and related acyclic alcohols. The Journal of Organic Chemistry. 1990;**55**:116-122

[34] Ute K, Hirose K, Kashimoto H, Hatada K, Vogl O. Haloaldehyde polymers. 51. Helix-sense reversal of isotactic chloral oligomers in solution. Journal of the American Chemical Society. 1991;**113**:6305-6306. DOI: 10.1021/ja00016a076

[35] Ute K, Oka K, Okamoto Y, Hatada K, Xi F, Vogl O. Haloaldehyde polymers LIII. Optical resolution of purely isotactic oligomers of chloral: Optical activity of the chloral oligomers assuming one-handed helical conformation in solution. Polymer Journal. 1991;**23**:1419-1424

[36] Tian G, Lu Y, Novak BM. Helix-sense selective polymerization of carbodiimides: Building permanently optically active polymers from achiral monomers. Journal of the American Chemical Society. 2004;**126**:4082-4083

[37] Schlitzer DS, Novak BM. Trapped kinetic states, chiral amplification and molecular chaperoning in synthetic polymers: Chiral induction in Polyguanidines through ion pair interactions. Journal of the American Chemical Society. 1998;**120**(9):2196-2197

[38] Tang H-Z, Lu Y, Tian G, Capracotta MD, Novak BM. Stable helical polyguanidines: Poly{N-(1-anthryl)-N‘-[(R)-and/or (S)-3,7-dimethyloctyl]guanidines}. Journal of the American Chemical Society. 2004;**126**:3722-3723

[39] Muller M, Zentel R. Interplay of chiral side chains and helical Main chains in polyisocyanates. Macromolecules. 1996;**29**:1609-1617

[40] Freire F, Seco JM, Quiñoá E, Riguera R. The prediction of the absolute stereochemistry of primary and secondary 1,2-Diols by 1H NMR spectroscopy: Principles and applications. Chemistry–A European Journal. 2005;**11**(19):5509-5522

[41] Arias S, Freire F, Quiñoá E, Riguera R. Nanospheres, nanotubes, toroids, and gels with controlled macroscopic chirality. Angewandte Chemie International Edition. 2014;**53**(50):13720-13724

[42] Materials Studio Modeling v.2018, BIOVIA Cambridge, U.K. Inc.; 2018

Chapter 4

Preparation, Characterization, and Applications of Carbonaceous Mesophase: A Review

Guanming Yuan and Zhengwei Cui

Abstract

Carbonaceous mesophase with a nematic liquid crystal structure possesses an easily graphitizable characteristic and can be used as a promising raw material to prepare anisotropic carbon and graphite materials with high performance and multifunction. Therefore, the carbonaceous mesophase occupies a pivotal and irreplaceable position in many frontier and cutting-edge fields. The controllable preparation and characterization of carbonaceous mesophase derived from a model molecule (i.e., naphthalene) are presented, especially the formation, development, and transformation of anisotropic liquid crystalline mesophase in the synthetic naphthalene pitch during the process of liquid-phase carbonization (350–450°C). The increasing applications of naphthalene-based carbonaceous mesophase as an ideal precursor material for fabricating representative advanced carbon materials with high added value (e.g., mesophase pitch-derived coke, mesocarbon microbeads, mesophase pitch-based carbon foam, high-modulus mesophase pitch-based carbon fibers, and high-thermal-conductivity carbon-based composites, etc.) are reviewed in detail in this chapter.

Keywords: carbonaceous mesophase, liquid crystal, carbon materials, preparation, characterization, applications

1. Introduction

The research on carbonaceous mesophase can be traced back to the 1960s, when Books and Taylor found there was liquid-crystalline phase (i.e., mesophase spheres) in the thermal conversion of carbonaceous feedstocks, which opens a new era in the research of liquid-phase carbonization and the development of carbon material industry [1]. Up to now, carbonaceous mesophase has been studied for more than 50 years and has always been the research hotspot and focus in the field of carbon materials [2–5]. Mesophase pitch has long been recognized as a liquid crystal in a defined temperature range (e.g., 200–400°C) and exhibits both lyotropic and thermotropic nature, which is different from ordinary polymers and isotropic pitch [2]. It is well known that the mesophase pitch with a nematic liquid crystal structure possesses an easily graphitizable characteristic and can be preferentially aligned under mechanical force shearing after melting; thereby it is regarded as a basic raw material for preparing high-performance carbon and graphite materials with controllable structure of forming an ordered graphite, which provides a feasible route to prepare graphite-like materials [2]. In addition to the high carbon yield and potential price advantage (owing to the relatively low cost of carbonaceous raw materials) of

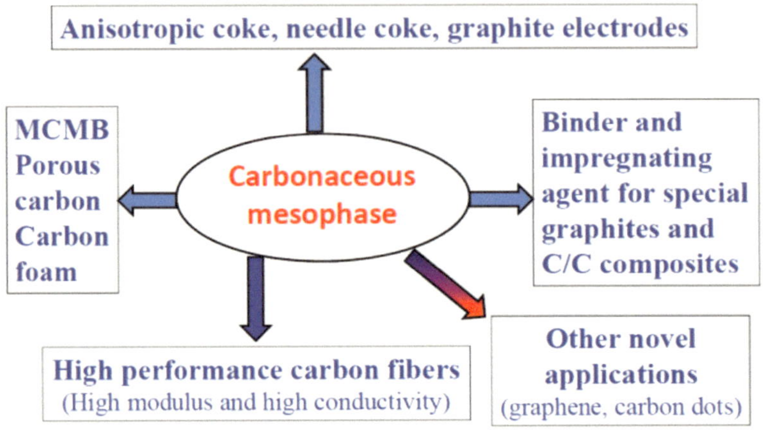

Figure 1.
Main promising applications of carbonaceous mesophase as an excellent precursor for making a wide variety of industrial and engineering carbon products.

mesophase pitch, it has become a high-quality precursor material for fabricating high-performance and multifunctional carbon materials as shown in **Figure 1**, such as mesophase pitch-based coke, needle coke, high-power graphite electrodes, mesocarbon microbeads (MCMBs), mesophase pitch-based carbon foam, mesophase pitch-based carbon fibers with high modulus and thermal conductivity, good binder and impregnating agent for high-thermal conductivity carbon-based composites, etc. Therefore, there is no doubt that mesophase pitch occupies a pivotal and irreplaceable position in various fields, such as defense, military, aerospace, cutting-edge technology, high-end industrial manufacturing, etc. [2, 3, 5].

2. Preparation of carbonaceous mesophase

The formation of mesophase pitch is a phase inversion process (transformed from isotropic to anisotropic), which is a result in which the pitch precursor undergoes thermal decomposition and thermal polycondensation to a certain extent. Nowadays, either thermal polycondensation of commercial coal-tar pitch and petroleum pitch (or even their certain soluble fractions) or catalytic polymerization of some aromatic substances is commonly used to prepare the carbonaceous mesophase [5]. It needs to be emphasized that carbonaceous precursors for the preparation of high-performance mesophase pitch are very crucial. Although commercial coal-tar pitch or petroleum pitch or heavy oil is very cheap and easy to obtain, these feedstocks are normally a complex mixture including with some heteroatoms and inorganic ash (~0.2 wt.%), which makes a spinnable mesophase pitch difficult to prepare [6–8]. So far, it is still very hard to massively produce cheap mesophase pitch with high quality, especially for continuously melt spinning high-performance carbon fibers. In recent 30 years, small model aromatic compounds (such as naphthalene, methylnaphthalene, anthracene, etc.) have been widely used to synthesize spinnable mesophase pitch by catalyzing with superacid, $HF\text{-}BF_3$. The obtained naphthalene-derived mesophase pitch possesses characteristics of high purity, controllable molecular structure, and ideal physical property [3, 9–11]. However, the severe corrosion problem and potential operating risk of using $HF\text{-}BF_3$ as a catalyst will unfortunately limit its widespread use (and such a mesophase pitch product named "AR" as shown in **Figure 2(a)** is now no longer available from, e.g., Mitsubishi Gas Chemical Company).

Figure 2.
Optical photographs of (a) pellet and (b) block-shaped carbonaceous mesophase pitch derived from naphthalene.

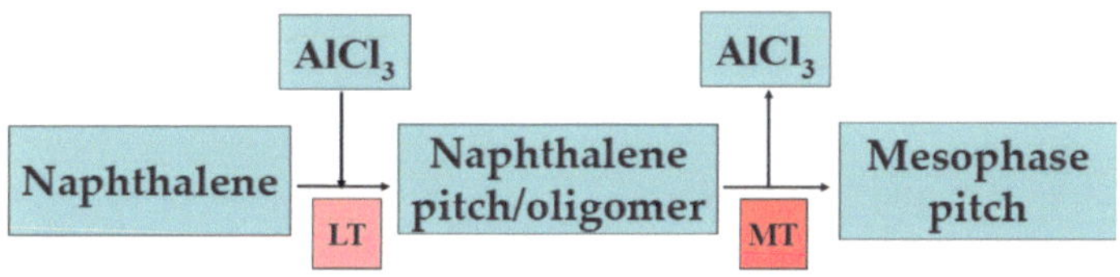

Figure 3.
Flow diagram of catalytic thermal polymerization of naphthalene molecule to form carbonaceous mesophase pitch ((LT) a low temperature of ~200°C and (MT) a mid temperature of ~430°C).

In the meantime, a mild catalyst $AlCl_3$ has been selectively used to prepare the mesophase pitch from the simple molecules and achieve the anticipated catalytic polymerization effect in spite of a trace of residual catalyst (e.g., 300–1000 ppm) inevitably intermingled in the mesophase pitch [3, 12–14]. **Figure 3** shows the flow diagram of catalytic thermal polymerization of naphthalene molecule to prepare carbonaceous mesophase pitch as shown in **Figure 2(b)** by a two-step reaction process at a liquid-phase carbonization temperature of 350–450°C for a certain period of time. It could be concluded that some carbonaceous precursors (e.g., naphthalene) have undergone four stages of liquid-crystalline sphere development and transformation and finally formed a bulk liquid-crystalline mesophase from an isotropic matrix as illustrated in **Figure 4** under a suitable reaction condition (i.e., reaction temperature and time) [2, 15, 16]. The general four-stage conversion of liquid crystals during the whole process is diagramed as follows: (I) generation of optically anisotropic spheres in isotropic matrix, (II) growth of anisotropic spheres in isotropic matrix, (III) coalescence of anisotropic spheres in isotropic matrix, and (IV) deformation and disintegration of anisotropic coalesced spheres to form bulk liquid-crystalline mesophase.

However, it has been demonstrated that the formation, development, and transformation of liquid-crystalline anisotropic spheres (i.e., nucleation, growth, coalescence and deformation and orientation) in an isotropic pitch matrix are unconcerted and inhomogeneous during the process of liquid-phase carbonization as shown in **Figure 5** [16]. Furthermore, it is not easy to obtain a 100 vol.% anisotropic mesophase pitch (i.e., bulk mesophase) both with a fine flow optical texture and an acceptable softening point less than 300°C for subsequent fiber spinning. This mainly depends on the carbonaceous precursors (e.g., molecular unit size, the flatness of molecules and the chemical reactivity, etc.) and the suitable thermal reaction conditions adopted [2, 15, 16].

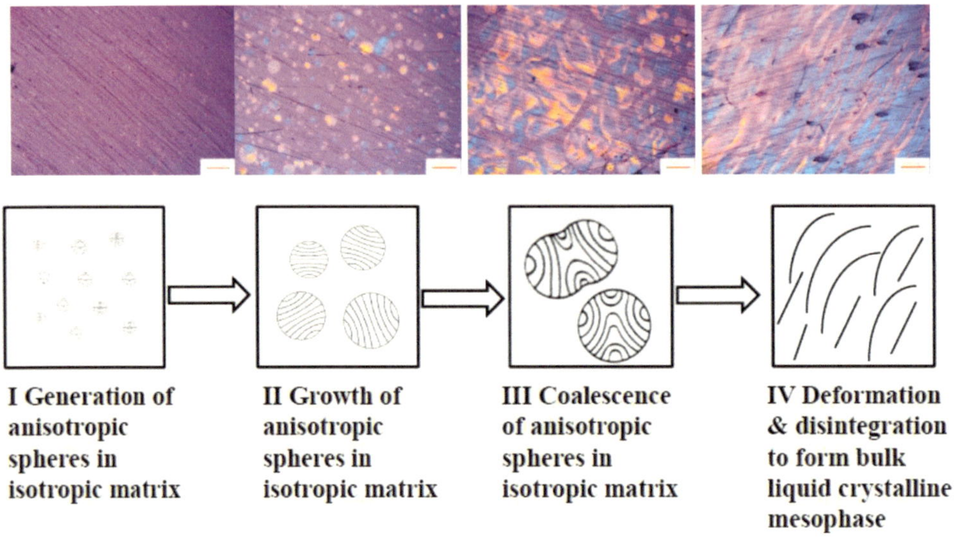

Figure 4.
Schematic illustration of the formation and development process of bulk liquid-crystalline mesophase under a suitable reaction condition (scale bar in PLM micrographs is 100 μm).

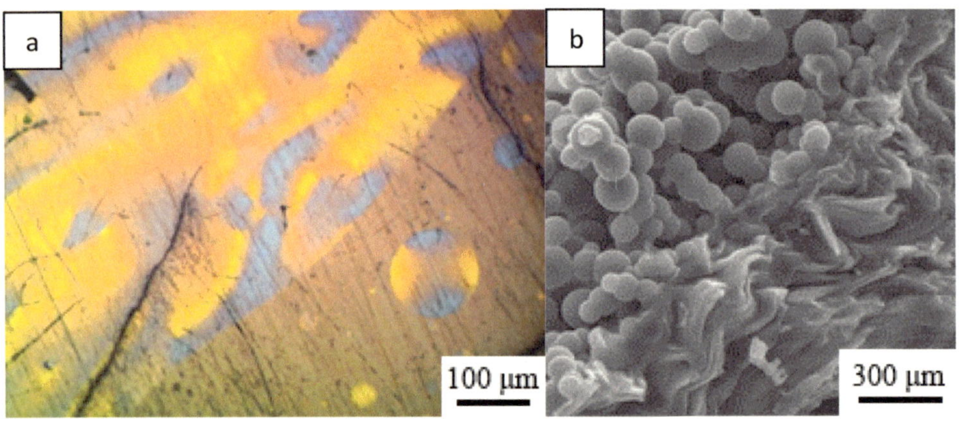

Figure 5.
(a) Polarized light microscope (PLM) micrograph of the naphthalene-based synthetic pitch and (b) SEM image of broken surface of the pitch-derived coke showing an unsynchronized and inhomogeneous conversion of liquid-crystalline anisotropic spheres.

3. Characterization of carbonaceous mesophase

Mesophase pitch consists of a large variety of polycyclic aromatic hydrocarbons and maintains the molecular ordering (i.e., optical anisotropy), which is an important precursor for high-performance industrial carbon materials. Characterizing the structures and properties of carbonaceous mesophase plays a significant role in its quality control, process optimization, and applications [5, 17, 18]. Only through effective measurement of the molecular weight distribution and quantitative description of the structural characteristic as well as the multi-scale evaluation of the thermophysical nature will the understanding, controllable preparation, and applications of carbonaceous mesophase be updated. The common instruments used for characterizing carbonaceous mesophase are as follows: Fourier-transform infrared spectrometer (FTIR), elemental analyzer, nuclear magnetic resonance

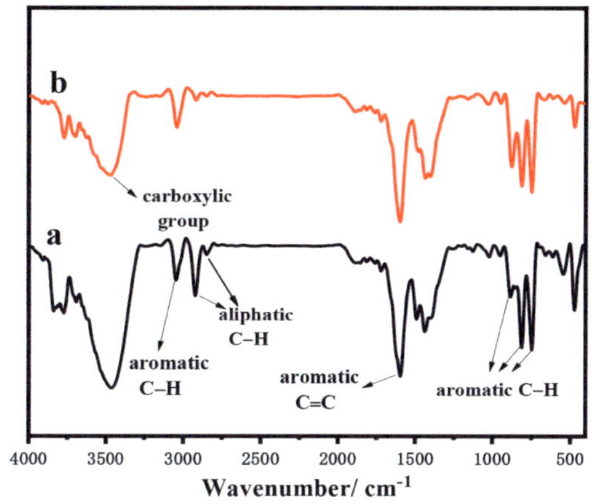

Figure 6.
FTIR patterns of (a) naphthalene pitch and (b) its derived mesophase pitch.

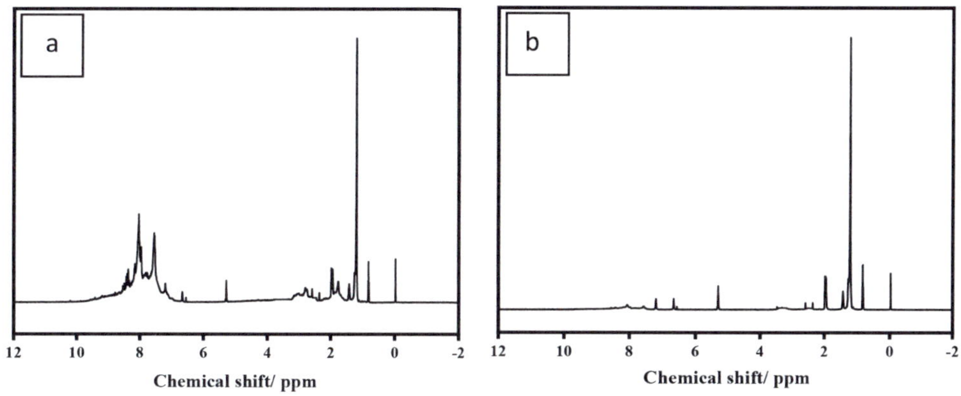

Figure 7.
^{1}H-NMR spectra of the soluble fractions from (a) naphthalene pitch and (b) its derived mesophase pitch.

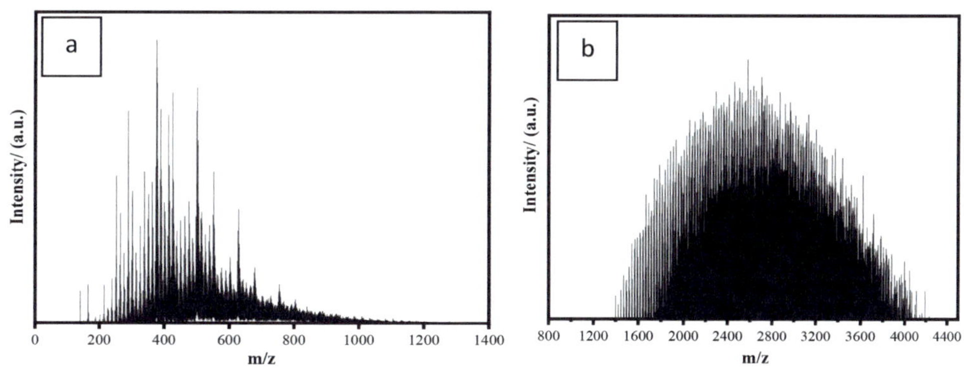

Figure 8.
MS spectra of (a) naphthalene pitch and (b) its derived mesophase pitch.

(NMR), flight mass spectrometer (MS), polarized light microscope, capillary rheometer, X-ray diffractometer, Raman spectrum, thermogravimetric, differential scanning calorimetry, etc.

The carbonaceous mesophase pitch prepared by $AlCl_3$ catalytic thermal polymerization of naphthalene has a relatively high aromaticity (the aromatic index is about 0.70)

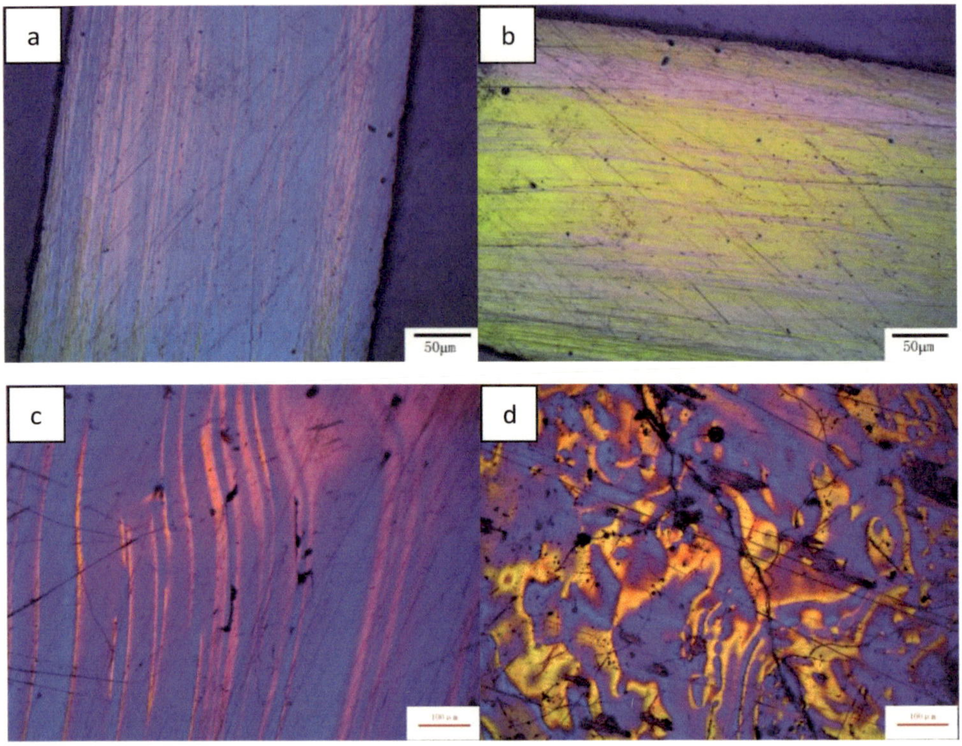

Figure 9.
Typical PLM micrographs of the (a, b) as-received, (c) melted, and (d) melt-stirred naphthalene-based AR mesophase pitch.

and a regular planar molecular structure constructed by a number of naphthenic structure, as well as a relatively large molecular weight of ~2600 g/mol, consisting of mesogen units (a ladder-shaped molecular structure) formed by ~20 naphthalene molecules through thermally induced aromatic growth [3, 10, 11] according to the analyses of **Figures 6–8**. The suitable softening point (260–280°C) and appropriate H/C mol ratio (0.52–0.60), as well as high liquid-crystalline mesophase content (100 vol.%) and ideal fine flow texture as displayed in **Figures 9** and **11**, are the significant characteristics of such carbonaceous mesophase. The analysis results of other characterizations are not shown here (refer to previous work [2, 3, 10, 11, 15, 18]).

4. Texture and properties of carbonaceous mesophase as a liquid crystal

The as-received liquid-crystalline AR mesophase pitch as shown in **Figure 9(a, b)** possesses a streamline "fibrous" texture with highly preferred orientation visible via orthogonal observation by rotating the object stage of the PLM. Following melting and melt-stirred treatments at 320°C as shown in **Figure 9(c, d)**, respectively, the optical texture of the melting pitch is nearly maintained, and the conformation and orientation of the macromolecules in the melt-stirred pitch are disrupted to become partially disordered or turbulent (severely deformed) depending upon the degree of stirring [19]. The purpose of this thermo-stirring treatment is to investigate the influence of liquid-crystalline texture of mesophase pitch precursors on the morphology, microstructure, and physical properties of resulting carbon fibers as shown in **Figure 10**.

It can be found that the as-prepared naphthalene-based mesophase pitch as being transmitted from the reaction autoclave to a metal plate at a molten status

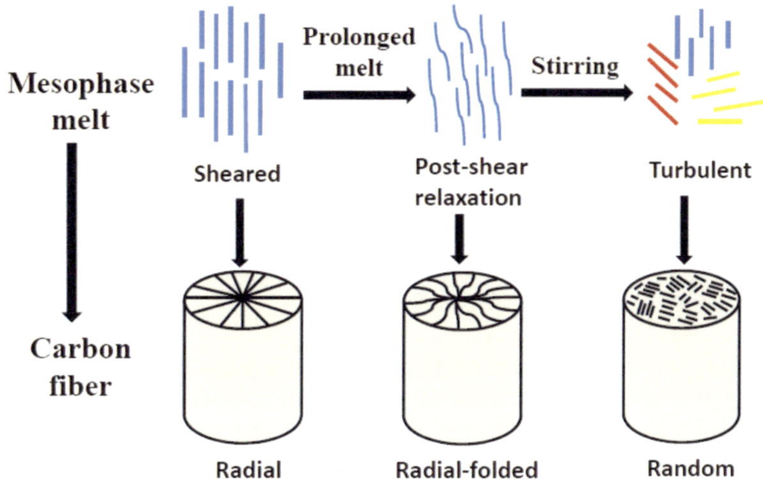

Figure 10.
Schematic of the microstructure evolution from mesophase pitch precursor to transverse texture-controlled carbon fibers as degree of melt-stirring increases.

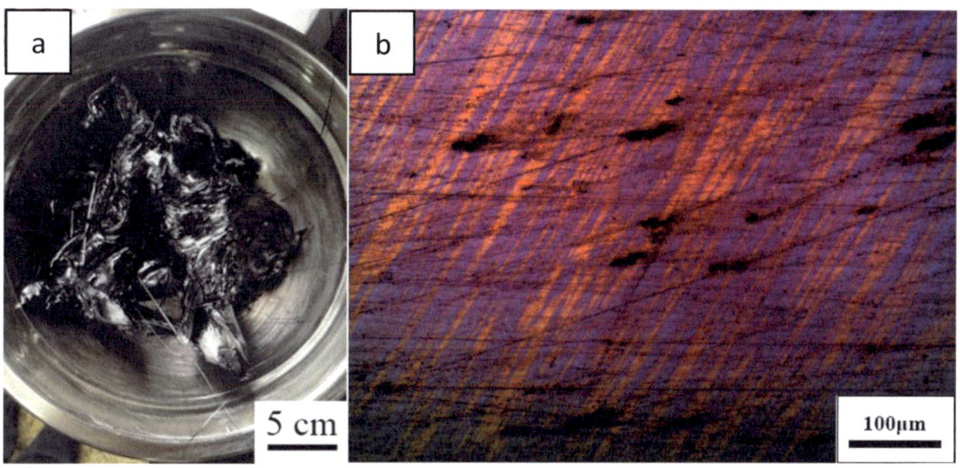

Figure 11.
(a) Optical photograph of a naphthalene-based synthetic pitch with good wire-drawing performance and (b) PLM micrograph of the drawn pitch fiber.

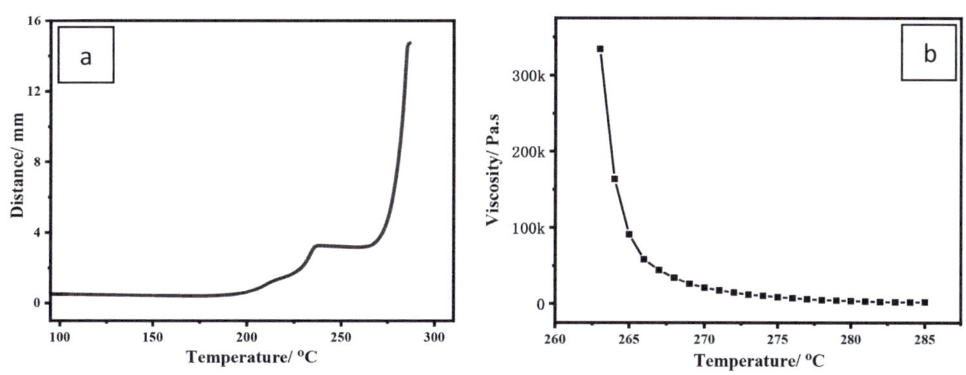

Figure 12.
Typical (a) molten flow curve of distance-temperature and (b) viscosity-temperature curve of naphthalene-based AR mesophase pitch.

exhibits good wire-drawing performance and ideal viscoelastic property and the unwittingly drawn wires (i.e., large-diameter pitch fibers) possess an orderly liquid-crystalline texture as shown in **Figure 11**, which is closely related to its plastic flowing behavior and low apparent viscosity upon melting as shown in **Figure 12**. This is favorable for pitch melt spinning and other rheology applications [5].

5. Applications of carbonaceous mesophase

5.1 Carbonaceous mesophase-derived coke

It is well known that pitch-derived coke is mainly used to make carbon and graphite electrodes equipped within electric arc furnaces for steelmaking, and mesophase pitch-derived coke (or needle coke) has an overwhelming advantage to produce graphite electrodes with high and ultrahigh power [5, 20].

It can be clearly seen that mesophase pitch-derived coke exhibits a well-oriented texture as shown in **Figure 13(a, b)**, which is closely related to the formation and development of flow-type liquid crystalline in carbonaceous mesophase products during the process of delayed coking [5]. In contrast, coarse-grained mosaic texture is presented in the coke derived from commercial coal-tar pitch as shown in **Figure 13(c, d)** [16, 20]. Thus it can be concluded that the carbonaceous feedstocks have a significant influence on the optical texture and microstructure of resulting coke, which depends on the development and evolution of carbonaceous mesophase during the liquid-phase carbonization process.

Figure 13.
(a, c) Optical photographs and (b, d) PLM micrographs of mesophase pitch-derived coke (a, b) and coal-tar pitch-derived coke (c, d).

5.2 Carbonaceous mesophase-based mesocarbon microbeads

As a special type of carbon material, MCMB has some outstanding physical and chemical properties that other carbon materials do not have due to its unique spherical morphology and lamellar structure. Therefore, MCMB can be widely applied to various fields, such as high-performance liquid chromatography column materials, high-specific surface area activated carbon materials, high-efficiency lithium ion battery anodes, high-density and high-strength graphite materials, etc. [5, 21].

Under suitable thermal reaction conditions, homogeneous liquid-crystalline spheres with an identical diameter of ~10 μm which appeared in the optically isotropic pitch matrix can be achieved as shown in **Figure 14(a)**, which is closely related to the effective control of the polymerization degree of naphthalene molecules. Through subsequent separation, infusibilization, and carbonization treatments, uniform-sized MCMBs as shown in **Figure 14(b)** can be easily obtained by starting with a simple naphthalene molecule.

5.3 Carbonaceous mesophase-based porous carbon and carbon foam

Recently, many researchers have used mesophase pitch as a raw material to prepare porous carbon materials (e.g., ultrahigh surface area activated carbon, mesoporous carbon, and hierarchical porous carbon) with controlled microstructure and morphology [22, 23]. The large specific surface area, rich pore structure and excellent adsorption performance of porous carbon materials provide excellent supporting characteristics for various transition metal and precious metal catalysts. Porous carbon support can resist the severe corrosion in harsh environments such as acid, alkali and salts, and greatly improve the adsorption performance and catalytic efficiency, and thus has broad applications [24].

Mesophase pitch-based carbon foam is a new type of porous carbon material prepared by foaming mesophase pitch as shown in **Figure 15**. Owing to its low density, high thermal and electrical conductivity, fire resistance, microwave absorption, noise reduction, low thermal expansion coefficient, chemical resistance, etc., carbon foam is extremely suitable for heat transfer systems, such as aerospace vehicles and satellites, rocket launching platforms, large heat exchangers, and computers in chemical plants [25–27]; therefore, such carbon foam sees promising application prospects.

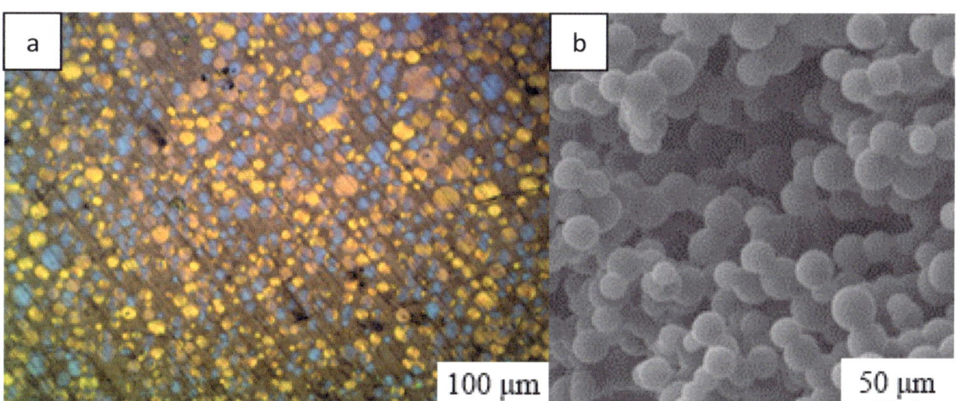

Figure 14.
(a) PLM micrograph of anisotropic liquid-crystalline carbonaceous spheres generated from naphthalene-based synthetic pitch and (b) SEM image of homogeneous MCMBs derived from the spherical liquid crystals.

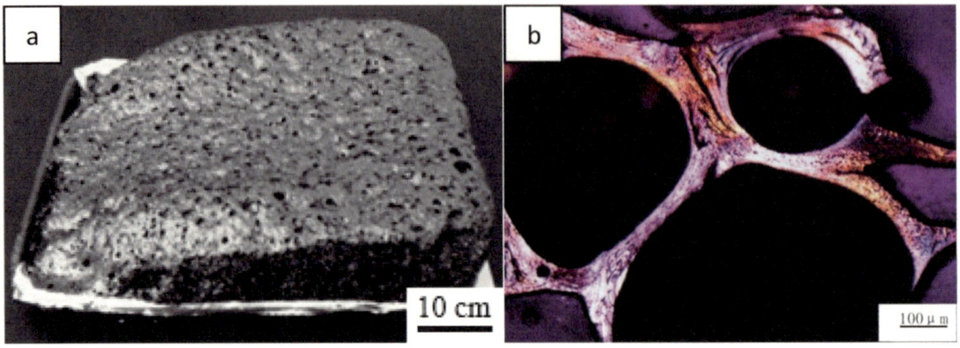

Figure 15.
(a) Optical photograph and (b) PLM micrograph of carbon foam derived from naphthalene-based mesophase pitch.

5.4 Carbonaceous mesophase-based carbon fibers

Mesophase pitch-based carbon fibers firstly reported by Singer in 1978 are the most successful high-end product for the development and application of carbonaceous mesophase, which are derived from spinnable mesophase pitch by melt spinning, oxidative stabilization, and carbonization and graphitization treatments [28]. The inherent alignment structure of liquid crystal molecules is preserved within the as-spun pitch fibers. Upon high-temperature graphitization, the graphite crystals are preferentially oriented along the fiber axis, so the final fibers have super high Young's modulus (up to a theoretical value of graphite, 1000 GPa) and excellent axial electrical (as low as 1.0 μΩ m in electrical resistivity) and thermal conductivity (exceeding 1000 W/m K). Thus they are now being widely used in aviation, aerospace, nuclear, and other high-tech fields, in which polyacrylonitrile-based carbon fibers have a certain limitation [3, 5, 29–33]. At present, only the United States (Cytec Industries Incorporated) and Japan (Mitsubishi Chemical Corporation and Nippon Graphite Fiber Corporation) have mature manufacturing technology ranging from the precursor materials to the final products (i.e., mesophase pitch, high-performance carbon fiber continuous filaments, and carbon fiber composites). The morphology of commercial carbon fibers usually includes three types of forms, i.e., continuous filament, chopped fiber, and ground fiber powder.

The round-shaped carbon fibers with different diameters and large-sized ribbon-shaped carbon fibers (sectional width ~2 mm, thickness ~10 μm) as shown in **Figure 16** can been successfully prepared from the AR mesophase pitch owing to its good spinnability. It is worthy to point out that most large-diameter carbon fibers with a radial transverse texture are inclined to spit in the subsequent high-temperature heat treatment. The ribbon-shaped carbon fibers can efficiently solve the crack problem and maintain their shape and structure without any damage. The carbon crystalline structure and layered orientation parallel to the ribbon main surface are obviously better than those of round fibers. The axial electrical resistivity and thermal conductivity of the round and ribbon fibers graphitized at 3000°C are measured to be as low as 1.1–1.30 μΩ m and about 900–1000 W/m K at room temperature [19, 34–36].

5.5 Carbonaceous mesophase-based carbon composites

Mesophase pitch-based carbon (graphite) fibers are often used as ideal functional fillers for preparing various carbon-based composites with high thermal conductivity [5, 37–41], which can be widely utilized in the field of thermal management [32, 33]. The thermal conductivity of these carbon-based composites depends not only on the

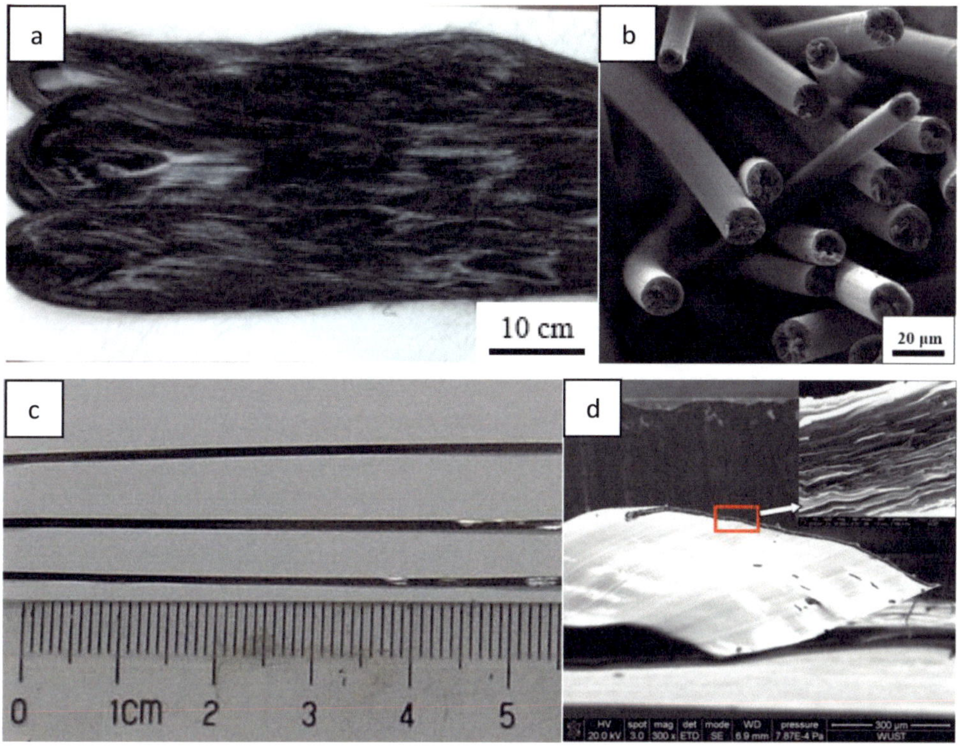

Figure 16.
(a, c) Optical photographs and (b, d) SEM micrographs of round- (a, b) and ribbon-shaped carbon fibers (c, d) derived from naphthalene-based AR mesophase pitch.

conduction performance of carbon fibers themselves and their loading amount, as well as laying or weaving architecture in the composites, but also on the physical properties of matrix materials involved (i.e., the resin, mesophase pitch, or pyrolytic carbon).

In the previous work, the mesophase pitch-based graphite fiber (long filament) reinforced one-dimensional (as shown in **Figure 17(a)**–**(c)**) and two-dimensional ABS resin composites with a large size of 10 cm × 10 cm x 0.3–2 cm can reach a high thermal conductivity of ~500 W/m K [37, 38]. However, the thermal conductivity of composites reinforced by shortcut carbon fibers and milled fiber powders as shown in **Figure 17(d, e)** is only 10–20 W/m K, which can be used as heat paste or thermal grease for interfacial heat dissipation. Using various mesophase pitch-based graphite fibers (i.e., round-shaped and ribbon-shaped fibers) as a reinforcing filler and the same mesophase pitch as a binder, ultrahigh thermal conductivity (700–900 W/m K) of the one-dimensional C/C composites as shown in **Figures 18** and **19** could be realized [39, 40]. However, it is disadvantage to use phenolic resin as a binder to prepare high-thermal-conductivity materials owing to its non-graphitizable nature (i.e., a typical hard carbon) as shown in **Figure 20**. By comparison, the mesophase pitch-derived carbon after high-temperature treatment exhibits good crystallinity, high graphitization degree, and orderly stacked graphene sheets as shown in **Figure 18(d)**, which is very important to improve the directional thermal conductivity performance. It is worth noting that the pyrolytic carbon with a highly oriented texture deposited on the mesophase pitch-based graphite fibers as shown in **Figure 21** is also found to markedly increase the thermal conductivity of C/C composites [41].

It is interesting to note that mesophase pitch is a promising binder (due to its good flow orientation performance in the molten state, easily graphitizable

Figure 17.
(a) Optical photograph, (b–d) PLM micrographs, and (e) SEM image of ABS resin composites reinforced by unidirectional (b, c) and disordered (d, e) mesophase pitch-based carbon fibers ((b, c) are, respectively, imaged perpendicular and parallel to the fiber axis).

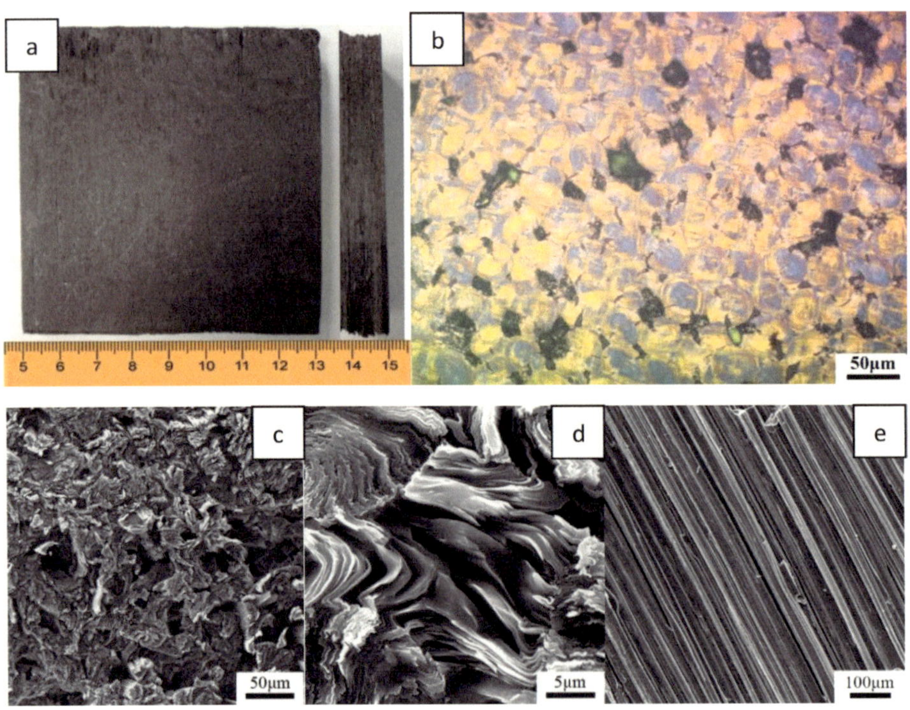

Figure 18.
(a) Optical photograph, (b) PLM micrograph, and (c)–(e) SEM images of unidirectional carbon/carbon composites reinforced by mesophase pitch-based carbon fibers using mesophase pitch as a binder ((b)–(d) are imaged perpendicular to the fiber axis, and (e) is imaged parallel to the fiber axis).

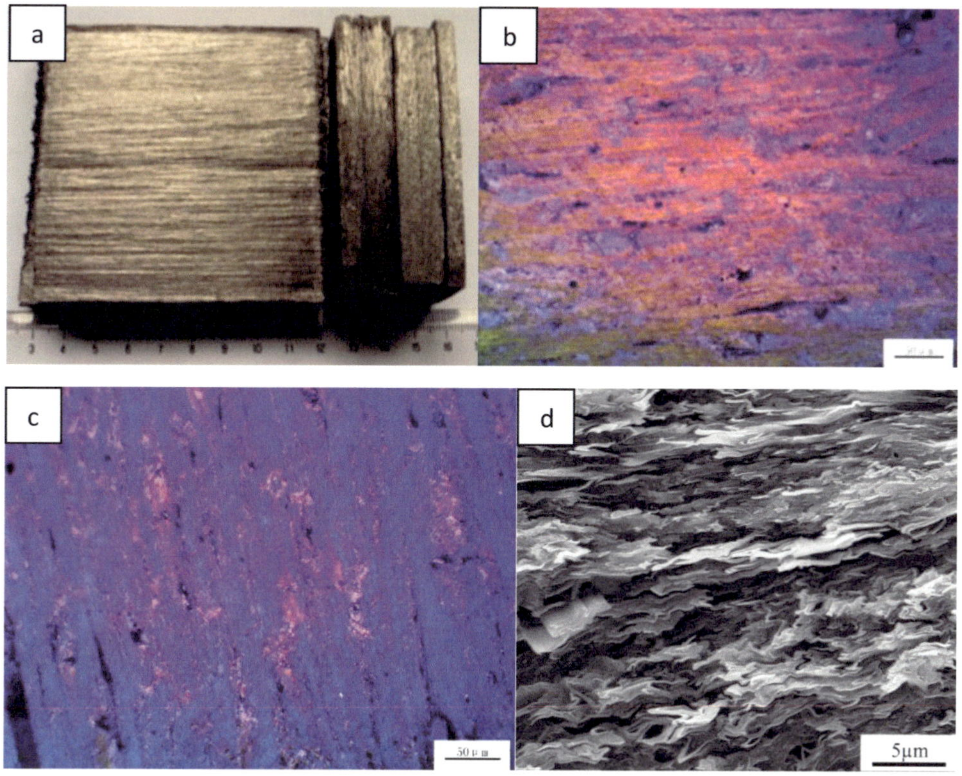

Figure 19.
(a) Optical photograph, (b, c) PLM orthogonal micrographs, and (d) SEM image of unidirectional carbon/carbon composites reinforced by mesophase pitch-based ribbon fibers using mesophase pitch as a binder.

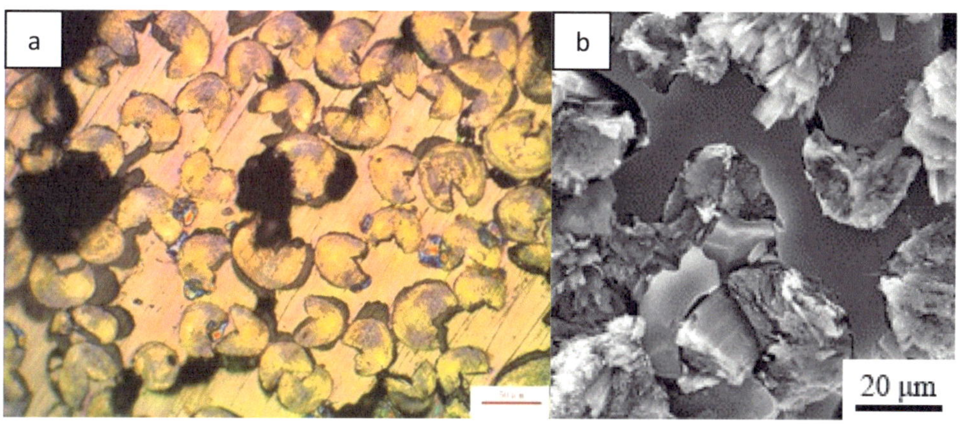

Figure 20.
(a) PLM micrograph and (b) SEM image of unidirectional carbon/carbon composites reinforced by mesophase pitch-based carbon fibers using phenolic resin as a binder.

characteristic, etc.) for large-scale fabricating natural flake graphite-molded blocks by using the cheap and available natural graphite flakes as a raw material. The prepared graphite blocks with a high bulk density of 1.9 g/cm^3 possess a highly preferred structural orientation perpendicular to the hot-pressing direction as shown in **Figure 22** and a high thermal conductivity of 500–600 W/m K in plane two-dimensional direction [42, 43].

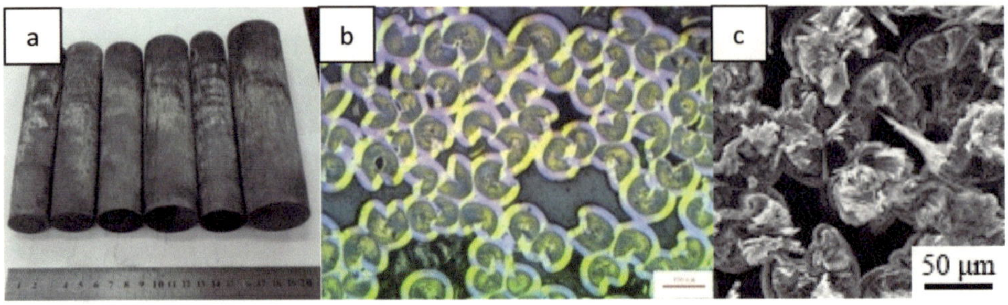

Figure 21.
(a) Optical photograph, (b) PLM micrograph, and (c) SEM image of unidirectional carbon/carbon composites reinforced by mesophase pitch-based carbon fibers using pyrolytic carbon as a "binder."

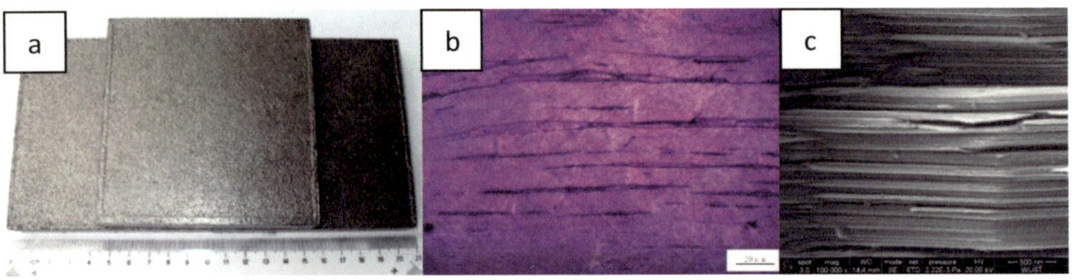

Figure 22.
(a) Optical photograph, (b) PLM micrograph, and (c) SEM image of natural flake graphite-molded blocks perpendicular to the hot-pressing direction using mesophase pitch as a binder.

5.6 Other applications

It is well known that carbon materials are important materials for the preparation of various batteries. From ancient dry batteries to today's high-efficiency fuel cells, as well as new high-energy storage batteries being developed, pitch-based carbon materials are playing an increasingly important role. Mesophase pitch is an easily graphitizable carbonaceous precursor. After high-temperature heat treatment, its three-dimensional stack structure is very regular, and mesophase pitch can be transformed into a high-crystalline graphite. The necessary energy of intercalating lithium ions into the carbon layers is relatively low, and thus such material has a large lithium insertion depth and reversible capacity [44, 45], especially carbonaceous mesophase-derived coke after spheroidizing and coating treatments as shown in **Figure 23(a)** which can significantly improve the cycle stability and service life of the battery.

By the same token, using the easily graphitized mesophase pitch-derived carbon as a raw material, a large-sized graphene (or a few layers of graphene sheets) with uniform thickness and good transparency as shown in **Figure 23(b)** can be successfully prepared through a special technique (i.e., molten salt ion intercalation stripping), which can realize the size and thickness control of carbon layers. The preparation method seems to be very simple and easy to operate and thus will have a good prospect.

In addition to being used as a high-quality raw material for the above-mentioned carbon materials, carbonaceous mesophase can also be used to prepare some novel and value-added carbon materials such as miracle graphene [46], carbon quantum dots [47], good binder for high-performance magnesia carbon bricks [48], fluorinated pitch [49], etc.

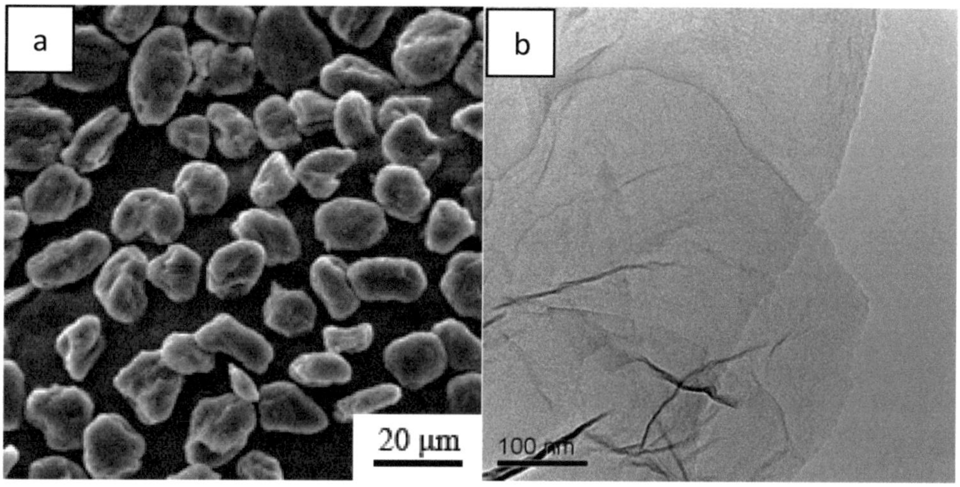

Figure 23.
Typical (a) SEM image of carbonaceous mesophase-derived spherical coke used as Li-ion battery anodes and (b) TEM image of carbonaceous mesophase-derived graphene with a relatively large size.

6. Conclusions

In this chapter, the preparation, characterization, and applications of naphthalene-based carbonaceous mesophase are reviewed. With the continuous advancement of preparation techniques and characterization methods, the understanding of the molecular structure, molecular weight, molecular weight distribution, aggregation texture, and rheology property of mesophase liquid crystals will be deepened, and finally the comprehensive understanding of the carbonaceous mesophase (including the formation mechanism, molecular dynamic law and high-efficiency control) from molecular and micro and macro scales could be realized, which will maximize the performance of carbonaceous mesophase-derived carbon products with desirable performance, multi-versatility, and high added value, thus to promote the theoretical foundation of carbonaceous mesophase and accelerate its broad applications in various fields.

Acknowledgements

The authors would like to thank professor Xuanke Li for his good suggestion and professional advice. This work was supported by the National Natural Science Foundation of China (Grant Nos. 91016003 and 51372177), the Hubei Provincial Department of Education Science Research Project (Grant No. Q20141104), the Key Laboratory of Hubei Province for Coal Conversion and New Carbon Materials (Grant No. WKDM201701), and the China Scholarship Council Fund (201808420114).

Conflict of interest

The authors declare no conflict of interest.

Author details

Guanming Yuan[1,2*] and Zhengwei Cui[1,2]

1 The State Key Laboratory of Refractories and Metallurgy, Wuhan University of Science and Technology, Wuhan, China

2 Hubei Province Key Laboratory of Coal Conversion and New Carbon Materials, Wuhan University of Science and Technology, Wuhan, China

*Address all correspondence to: yuanguanming@wust.edu.cn

References

[1] Brooks JD, Taylor GH. The formation of graphitizing carbons from the liquid phase. Carbon. 1965;**3**(2):185-193. DOI: 10.1016/0008-6223(65)90047-3

[2] Marsh H, Diez MA. Mesophase of graphitizable carbons. In: Shibaev VP, Lam L, editors. Liquid Crystalline and Mesomorphic Polymers. New York: Springer-Verlag New York Inc; 1994. pp. 231-257. DOI: 10.1007/978-1-4613-8333-8-7

[3] Mochida I, Korai Y, Ku CH, et al. Chemistry of synthesis, structure, preparation and application of aromatic-derived mesophase pitch. Carbon. 2000;**38**(2):305-328. DOI: 10.1016/S0008-6223(99)00176-1

[4] Hurt RH, Chen ZY. Liquid crystals and carbon materials. Physics Today. 2000;**53**(3):39-44. DOI: 10.1063/1.883020

[5] Wang CY. Theory and application of carbonaceous mesophase. Beijing PRC: Science Press; 2015. ISBN: 9787030457509

[6] Castro LDD. Anisotropy and mesophase formation towards carbon fibre production from coal tar and petroleum pitches—A review. Journal of the Brazilian Chemical Society. 2006;**17**(6):1096-1108. DOI: 10.1590/S0103-50532006000600006

[7] Li M, Zhang YD, Yu ST, et al. Preparation and characterization of petroleum-based mesophase pitch by thermal condensation with in-process hydrogenation. RSC Advances. 2018;**8**:30230-30238. DOI: 10.1039/C8RA04679D

[8] Bermudez V, Lukubira S, Ogale AA. Pitch precursor-based carbon fibers. In: Zweben CH, Beaumont PWR, editors. Comprehensive Composite Materials II. 2nd ed. Amsterdam, Netherlands: Elsevier Science Ltd; 2018. pp. 41-65. DOI: 10.1016/B978-0-12-803581-8.10312-1

[9] Mochida I, Shimizu K, Korai Y, et al. Preparation of mesophase pitch from aromatic hydrocarbons by the aid of HF/BF_3. Carbon. 1990;**28**(2-3):311-319. DOI: 10.1016/0008-6223(90)90005-J

[10] Mochida I. Recent progresses of mesophase pitch a review at a receipt of 1993 Charles E. Pettinos Award, American Carbon Society. TANSO. 1994;**163**:150-162. DOI: 10.7209/tanso.1994.150

[11] Mochida I, Yoon SH, Korai Y. Mesoscopic structure and properties of liquid crystalline mesophase pitch and its transformation into carbon fiber. The Chemical Record. 2002;**2**(2):81-101. DOI: 10.1002/tcr.10016

[12] Boero JFR, Wargon JA. Study of the AlCl3 catalytic activity on aromatic hydrocarbons-II: Mesophase formation. Carbon. 1981;**19**(5):341-346. DOI: 10.1016/0008-6223(81)90057-9

[13] Mochida I, Sone Y, Korai Y. Preparation and properties of carbonaceous mesophase-II highly soluble mesophase from ethylene tar modified using aluminum chloride as a catalyst. Carbon. 1985;**23**(2):175-178. DOI: 10.1016/0008-6223(85)90009-0

[14] Kumar S, Srivastava M. Catalyzing mesophase formation by transition metals. Journal of Analytical and Applied Pyrolysis. 2015;**112**:192-200. DOI: 10.1016/j.jaap.2015.01.029

[15] Marsh H, Latham CC. The chemistry of mesophase formation. In: Bacha JD, Newman JW, White JL, editors. Petroleum-Derived Carbons. ACS Symposium Series. Washington DC: American Chemical Society; 1986.

pp. 1-28. DOI: 10.1021/bk-1986-0303.ch001

[16] Yuan GM, Jin Z, Zuo XH, et al. Effect of carbonaceous precursors on the structure of mesophase pitches and their derived cokes. Energy & Fuels. 2018;**32**(8):8329-8339. DOI: 10.1021/acs.energyfuels.8b01824

[17] Duan CT, Zheng DF, Liu JQ, et al. Research progress on the characterization of mesophase pitch. New Carbon Materials. 2018;**33**(3):193-202

[18] Thies MC. Fractionation and characterization of carbonaceous pitch oligomers: understanding the building blocks for carbon materials. In: Naskar AK, Hoffman WP, editors. Polymer Precursor-Derived Carbon. ACS Symposium Series. Washington, D.C.: American Chemical Society; 2014. pp. 85-136. DOI: 10.1021/bk-2014-1173.ch005

[19] Yuan GM, Li BL, Li XK, et al. Effect of liquid crystalline texture of mesophase pitches on the structure and property of large diameter carbon fibers. ACS Omega. 2019;**4**(1):1095-1102. DOI: 10.1021/acsomega.8b03189

[20] Marsh H, Martínez-Escandell M, Rodríguez-Reinoso F. Semicokes from pitch pyrolysis: Mechanisms and kinetics. Carbon. 1999;**37**(3):363-390. DOI: 10.1016/S0008-6223(98)00205-X

[21] Zhang DK, Zhang LZ, Fang XL, et al. Enhancement of mesocarbon microbead (MCMB) preparation through supercritical fluid extraction and fractionation. Fuel. 2019;**237**:753-762. DOI: 10.1016/j.fuel.2018.10.054

[22] Qiao WM, Song Y, Hong SH, et al. Development of mesophase pitch derived mesoporous carbons through a commercially nanosized template. Langmuir. 2006;**22**(8):3791-3797. DOI: 10.1021/la052494p

[23] Adelhelm P, Yong-Sheng H, Chuenchom L, et al. Generation of hierarchical meso- and macroporous carbon from mesophase pitch by spinodal decomposition using polymer templates. Advanced Materials. 2007;**19**(22):4012-4017. DOI: 10.1002/adma.200700699

[24] Zhang PF, Zhu HY, Dai S. Porous carbon supports: recent advances with various morphologies and compositions. ChemCatChem. 2015;**7**(18):2788-2805. DOI: 10.1002/cctc.201500368

[25] Gallego NC, Klett JW. Carbon foams for thermal management. Carbon. 2003;**41**(7):1461-1466. DOI: 10.1016/S0008-6223(03)00091-5

[26] Nagel B, Pusz S, Trzebicka B. Review: Tailoring the properties of macroporous carbon foams. Journal of Materials Science. 2014;**49**(1):1-17. DOI: 10.1007/s10853-013-7678-x

[27] Inagaki M, Qiu JS, Guo QG. Carbon foam: Preparation and application. Carbon. 2015;**87**:128-152. DOI: 10.1016/j.carbon.2015.02.021

[28] Singer LS. The mesophase and high modulus carbon fibers form pitch. Carbon. 1978;**16**(6):409-415. DOI: 10.1016/0008-6223(78)90085-4

[29] Minus ML, Kumar S. The processing, properties, and structure of carbon fibers. Journal of Metals. 2005;**57**(2):52-58. DOI: 10.1007/s11837-005-0217-8

[30] Frank E, Ingildeev D, Steudle LM, et al. Carbon fibers: precursor systems, processing, structure and properties. Angewandte Chemie, International Edition. 2014;**53**(21):5262-5298. DOI: 10.1002/anie.201306129

[31] Emmerich FG. Young's modulus, thermal conductivity, electrical resistivity and coefficient of thermal expansion of mesophase pitch-based

carbons fibers. Carbon. 2014;**68**:274-293. DOI: 10.1016/j.carbon.2014.07.068

[32] Zweben C. Advances in composite materials for thermal management in electronic packaging. Journal of Metals. 1998;**50**(6):47-51. DOI: 10.1007/s11837-998-0128-6

[33] Feng W, Qin MM, Feng YY. Toward highly thermally conductive all-carbon composites: Structure control. Carbon. 2016;**109**:575-597. DOI: 10.1016/j.carbon.2016.08.059

[34] Yuan GM, Li XK, Dong ZJ, et al. The structure and properties of ribbon-shaped carbon fibers with high orientation. Carbon. 2014;**68**:426-439. DOI: 10.1016/j.carbon.2013.11.019

[35] Xiong XQ, Yuan GM, Li XK, et al. Preparation and characterization of ribbon-shaped mesophase pitch-based carbon fibers with different crystal orientations. Journal of Inorganic Materials. 2014;**29**(11):1186-1192. DOI: 10.15541/jim20140101

[36] Yuan GM, Li XK, Xiong XQ, et al. A comprehensive study on the oxidative stabilization of mesophase pitch-based tape-shaped thick fibers with oxygen. Carbon. 2017;**115**:59-76. DOI: 10.1016/j.carbon.2016.12.040

[37] Yi J, Yuan GM, Li XK, et al. Preparation and characterization of large diameter pitch based carbon fiber/ABS resin composites with high thermal conductivities. New Carbon Materials. 2014;**30**(1):63-70. DOI: 10.1016/j.carbon.2015.02.009

[38] Yuan GM, Li XK, Yi J, et al. Mesophase pitch-based graphite fiber-reinforced acrylonitrile butadiene styrene resin composites with high thermal conductivity. Carbon. 2015;**95**:1007-1019. DOI: 10.1016/j.carbon.2015.09.019

[39] Lin JF, Yuam GM, Li XK, et al. Preparation of 1D C/C composites with high thermal conductivity. Journal of Inorganic Materials. 2013;**28**(12):1338-1344. DOI: 10.3724/SP.J.1077.2013.13110

[40] Yuan GM, Li XK, Dong ZJ, et al. Pitch-based ribbon-shaped carbon-fiber-reinforced one-dimensional carbon/carbon composites with ultrahigh thermal conductivity. Carbon. 2014;**68**:413-425. DOI: 10.1016/j.carbon.2013.11.018

[41] Zhang X, Li XK, Yuan GM, et al. Large diameter pitch-based graphite fiber reinforced unidirectional carbon/carbon composites with high thermal conductivity densified by chemical vapor infiltration. Carbon. 2017;**114**:59-69. DOI: 10.1016/j.carbon.2016.11.080

[42] Yuan GM, Li XK, Dong ZJ, et al. Graphite blocks with preferred orientation and high thermal conductivity. Carbon. 2012;**50**(1):175-182. DOI: 10.1016/j.carbon.2011.08.017

[43] Yuan GM, Xue Z, Cui ZW, et al. Controlled preparation and thermal conductivity of highly oriented graphite blocks. Journal of Inorganic Materials. 2017;**32**(6):587-595. DOI: 10.15541/jim20160480

[44] Mochida I, Ku CH, Yoon SH, et al. Anodic performance and mechanism of mesophase-pitch-derived carbons in lithium ion batteries. Journal of Power Sources. 1998;**75**(2):214-222. DOI: 10.1016/S0378-7753(98)00101-3

[45] Zhang C, Huang ZJ, Lv W, et al. Carbon enables the practical use of lithium metal in a battery. Carbon. 2017;**123**:744-755. DOI: 10.1016/j.carbon.2017.08.027

[46] Barreda D, Pérez-Mas AM, Silvestre-Albero A, et al. Unusual flexibility of mesophase pitch-derived carbon materials: An approach to the synthesis of graphene. Carbon. 2017;**115**:539-545. DOI: 10.1016/j.carbon.2017.01.046

[47] Wang HB, Ning GQ, He X, et al. Carbon quantum dots derived by direct carbonization of carbonaceous microcrystals in mesophase pitch. Nanoscale. 2018;**10**(45):21492-21498. DOI: 10.1039/C8NR07385F

[48] Kanno K, Koike N, Korai Y, et al. Mesophase pitch and phenolic resin blends as binders for magnesia-graphite bricks. Carbon. 1999;**37**(2):195-201. DOI: 10.1016/S0008-6223(98)00152-3

[49] Zhang JC, Shi JL, Wu GP, et al. Changes in the structure and functional groups produced during the fluorination of mesophase microbeads. Carbon. 2011;**49**(5):1628-1634. DOI: 10.1016/j.carbon.2010.12.046

Section 2

Display Technology

Chapter 5

AMOLED Displays with In-Pixel Photodetector

Nikolaos Papadopoulos, Pawel Malinowski, Lynn Verschueren, Tung Huei Ke, Auke Jisk Kronemeijer, Jan Genoe, Wim Dehaene and Kris Myny

Abstract

The focus of this chapter is to consider additional functionalities beyond the regular display function of an active matrix organic light-emitting diode (AMOLED) display. We will discuss how to improve the resolution of the array with OLED lithography pushing to AR/VR standards. Also, the chapter will give an insight into pixel design and layout with a strong focus on high resolution, enabling open areas in pixels for additional functionalities. An example of such additional functionalities would be to include a photodetector in pixel, requiring the need to include in-panel TFT readout at the peripherals of the full-display sensor array for applications such as finger and palmprint sensing.

Keywords: AR, VR, OLED, readout, TFT, high resolution, extra functionalities, fingerprint, additional, functionalities, lithography, in-panel, high resolution, photodetector

1. Introduction

Active matrix OLED (AMOLED) displays are today's mainstream consumer displays available in various form factors, such as smart watches, mobile displays, and large area television. They are highly appealing because of their wide viewing angles, nice color saturation and great potential for curved, flexible and/or rollable format. There are several options for (flexible) backplane technologies based on thin-film transistors (TFTs), namely, metal-oxide TFTs (such as indium gallium zinc oxide or IGZO), low-temperature polycrystalline silicon (LTPS) TFTs, or a combination of IGZO and LTPS, more precisely LTPO or low-temperature polycrystalline silicon and metal oxide. All these technologies have their pros and cons. Among others, IGZO semiconductors are n-type only which is sufficient for a backplane driving an OLED but is less adequate for peripheral circuits. Another key important asset of IGZO is the ultralow source-drain leakage current due to the large bandgap of the semiconductor, enabling long retention times of data storing, i.e., pixels are not leaking. In contrary, LTPS has both n- and p-type devices and thus the capability of CMOS circuits. LTPS transistors can drive larger currents due to the intrinsically higher mobility, enabling complex in-pixel compensation schemes and peripheral circuits. However, with LTPS transistors the leakage will be larger. This is a key reason why LTPO has been developed: this technology combines the ultralow leakage current of IGZO and a p-type LTPS transistor resulting in a hybrid complementary technology. An OLED is an organic LED emitting light directly proportional to its forward current. Therefore it requires a current source as driver in the pixel. In many cases, this is achieved by placing a TFT in series with the OLED and driving it in saturation.

In this chapter, we will investigate the potential to embed additional functionalities in the display. Therefore, several strategies will be discussed focusing on improving the resolution of the current displays, by technology optimization introducing photolithography patterning of the OLED and by design evaluating external compensation vs. internal compensation. The extra space in the pixel, due to the combination of photolithography and simple pixel circuit, provides opportunities to include extra functions at the same original area. The focus in this book chapter is to add a photosensitive detector for fingerprint and palmprint readout.

2. Display resolution roadmap for various applications

Resolution (number of pixels) and pixel pitch (size and spacing of pixels) are two main parameters defining the architecture of the display arrays. The first, expressed typically in megapixels, is standardized by the content type, resulting in different generations of TVs: VGA, full HD, 4K, and, most recently, 8K. The latter, expressed typically in pixels per inch (ppi) or pixels per degree (ppd), is used as a benchmark for smartphones, with high-end models featuring densities in the range of 600 ppi. This is a value that gives a good enough image quality for hand-held devices, with the viewing distance of approximately 30 cm (1 foot). At the same time, future near-to-eye augmented reality/virtual reality (AR/VR) displays impose ultrahigh definition, as the pixel density needs to be beyond the pattern resolving capabilities of the human retina (30 cycles per degree) [1, 2]. The resolution should be maximized to provide the highest possible output within the eye box in any given point of the 180° field of view (FOV), also to enable foveated rendering.

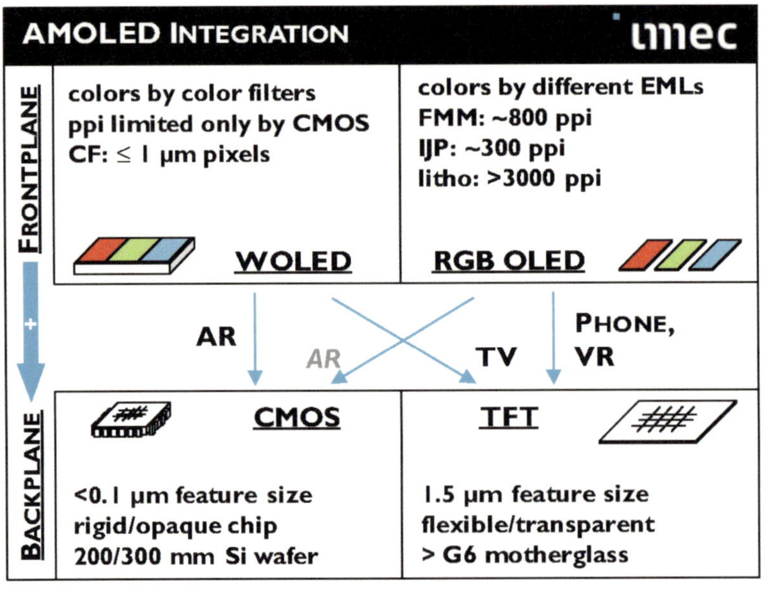

Figure 1.
Color-by-white vs. RGB OLED frontplane can be fabricated on top of CMOS or TFT backplane. Each combination is suitable for different applications.

An aperture ratio close to unity will eliminate the screen door effect and ensure natural experience. Transparency is necessary to avoid sense of isolation from the real-world view and to diversify from the virtual reality (VR) headsets (**Figure 1**). To realize all of the above, we need both the microdisplay-like pixel pitch down-scaling [3] and the flat-panel-display-like (FPD) backplane size up-scaling [4, 5]. Switching to advanced nodes in flat panel backplane manufacturing can result in ultrahigh-definition, direct-view AR displays fabricated in a cost-effective way.

3. Lithographic patterning of OLEDs to increase resolution

In order to realize an ultrahigh resolution display, all elements of the system (backplane, frontplane, and driving) need to provide appropriate pixel density. On the frontplane side, several options for the light source can be chosen (**Figure 2**). OLED technology currently dominates the smartphone display industry not only with performance but also with the cost structure. In this case, the colors are defined by depositing separate device stacks for each color, which is typically referred to as side-by-side, red-green-blue (RGB) array. In OLED TVs, one common white OLED stack is combined with a color filter array (CFA). The limitation of the side-by-side RGB array is the pixel density, limited by the fine metal masking (FMM) technology, which uses deposition through a metal mesh. The white OLED array can achieve very small pixel pitch, which is only limited by the backplane and CFA resolution but imposes brightness loss due to CF transmission. Patterning multicolor OLEDs by photolithography can address the needs of ultralow pixel pitch for the future AR displays by realizing side-by-side OLED stacks with extreme density.

Patterning OLEDs by photolithography is an emerging, disruptive fabrication technique. The main challenge is the extreme chemical sensitivity of OLED materials with solvent, moisture, air, and temperature exposure responsible for performance degradation. The choice of appropriate photolithography chemistry is crucial, with fluorinated [6] or non-fluorinated systems [7] as the dominant options. **Figure 3** shows the concept of using a negative-type photoresist to define patterns on top of OLED in a subtractive approach. First, the OLED stack is deposited as a plain layer over the entire substrate, on top of a pixel definition layer (PDL). This defines the active area of the light emitter. Second, photoresist is deposited on top of the entire substrate. Then, it is exposed through a lithography mask and developed to obtain the required pattern. Afterwards, the OLED layers that are not covered by the photoresist are etched away (typically with dry etching, such as reactive ion etch). In the end, the photoresist is stripped to achieve patterned OLED islands.

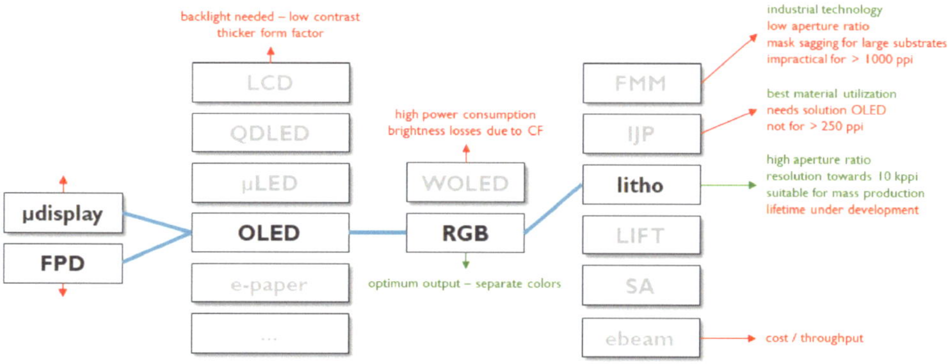

Figure 2.
Various display configuration options.

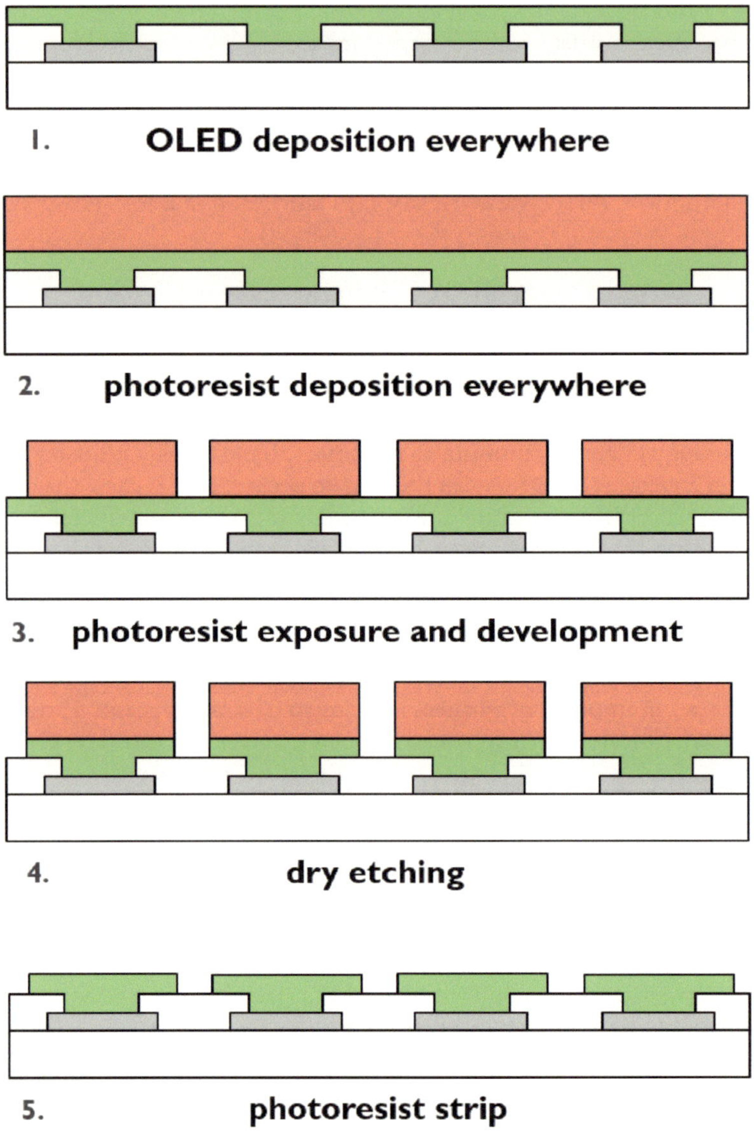

Figure 3.
Process flow for photolithography patterning of OLED stacks.

Photolithography allows pattern transfer beyond 1 μm resolution, enabling high-density lines and spaces. Transfer of small islands means that, with appropriate alignment (e.g., with an i-line stepper), a pixel density of a few thousand pixels per inch (ppi) can be realized. Transfer of openings means that pixel spacing can be minimized, resulting in a high aspect ratio. This is applicable for both TFT-based flat panel displays and CMOS-based microdisplays. Tests on patterning the OLED emission layer have shown that it is possible to achieve 1 μm pitch lines and spaces (**Figure 4**). Furthermore, the photoluminescence signal of the EML is maintained proving compatibility of this process with OLED material. 1 μm presented here is not a fundamental limit of the approach but rather a limit of the lithography mask design used in the experiment.

The achievable pixel density of the frontplane is limited not only by the photoresist used but also by the critical dimension (CD) and alignment/overlay accuracy of the litho tools used. In the i-line steppers typical for flat panel

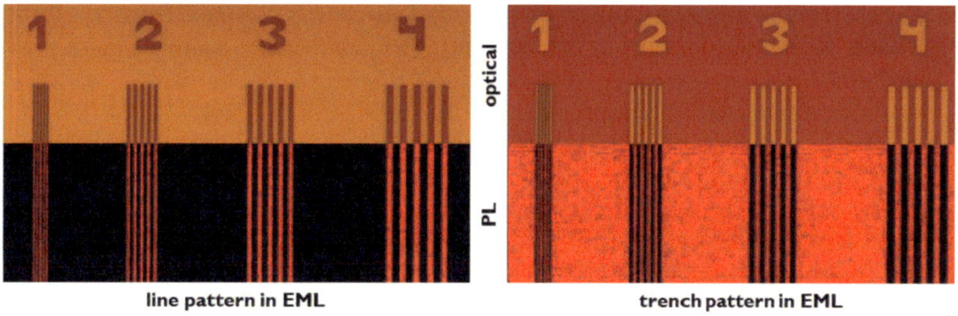

Figure 4.
OLED patterns of 1, 2, 3 and 4 µm: optical and corresponding photoluminescence pictures of red EML patterned as lines (left) and spaces (right).

manufacturing, the achievable CD is 1.5 µm with an overlay between 0.25 and 0.5 µm. In contrast, CMOS fabs used for microdisplay manufacturing feature more advanced semiconductor nodes, with 248 nm KrF or 193 nm ArF light sources. Assuming a minimum PDL opening (defining the active area) of 500 nm, a 1.5 µm node imposes a density limit of 3500 ppi (for RGB) with an aperture ratio below 5%. Going to KrF steppers, the achievable density increases to 10,000 ppi while keeping the aperture ratio above 35%. This demonstrates the need of a tooling upgrade for future AR displays, both for the frontplane and the backplane. Denser and more efficient packing of pixels requires scaling down of the technology node, especially in FPD manufacturing.

OLED patterning by photolithography means that the deposition of the stack is interrupted (vacuum break) and the photoresist interacts with the organic materials. In the most simple case, the photolithography process is performed in a clean room in ambient atmosphere. The devices are loaded back into the glove box after the etch step for each color and after the photoresist strip when all colors are finished. This raises a serious challenge for the device lifetime. If the process is not optimized for compatibility with the stack, the current-voltage-luminance (IVL) curve shifts to the right (increased turn-on voltage) and to the bottom (reduced luminance). As a consequence, the brightness of the patterned OLED drops very fast and disappears even after a few minutes (**Figure 5**). Optimization of the photoresist system, of the OLED stack [8] and of the fabrication process, is needed to achieve OLED performance enabling implementation into devices. At imec, we demonstrated phosphorescent green OLED with T90 lifetime of >150 h at the starting brightness of 1000 nit. Efficiency remained above 85 cd/A before and after patterning. Current performance is considered an important step on the path to industrial technology readiness level, estimated to be T97 of at least 1000 h (for the green stack) [8].

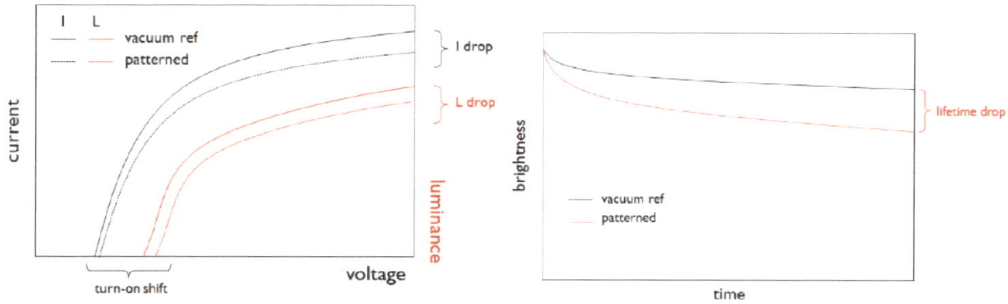

Figure 5.
IVL and lifetime curves for reference OLED and the possible effects of degradation by patterning.

Figure 6 shows an example comparison of unpatterned and patterned OLED lifetime curve at initial brightness of 1000 nit. The performance improvement can bring the two curves closer together.

OLED photolithography was used to fabricate passive displays with a 1400 × 1400 pixel array (almost 2 megapixels). 6 μm metal lines and 10 μm line pitch with SiN pixel definition layer (PDL) were used on glass substrate. Green and red OLED stacks were deposited by thermal evaporation in ultrahigh vacuum. After deposition of the first color (until above emission layer), photoresist was spin-coated, baked, exposed, and developed. Then, the OLED stack not covered by the photoresist was removed by dry etching. After that, the sample went back to the ultrahigh vacuum chamber for second color deposition, and the patterning process was repeated, this time finishing with stripping the photoresist. A semitransparent top contact stack was subsequently deposited, and the display was encapsulated with cavity glass. Both colors can be driven separately, and the PDL design allows for emission of a fixed image specified for each color (**Figure 7**). Subpixel pitch of 10 μm resulted in smooth edges and excellent feature representation. The device was tested for tens of hours with both colors on. No drop of brightness nor appearance of defects could be observed [9].

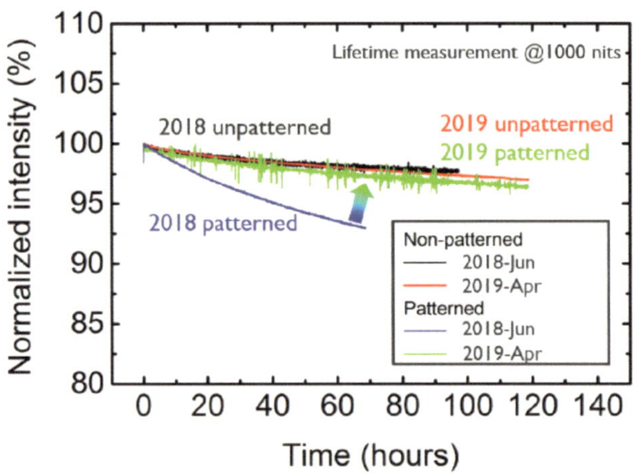

Figure 6.
Lifetime curves of a phosphorescent green OLED at 1000 nit starting brightness for unpatterned and patterned stack.

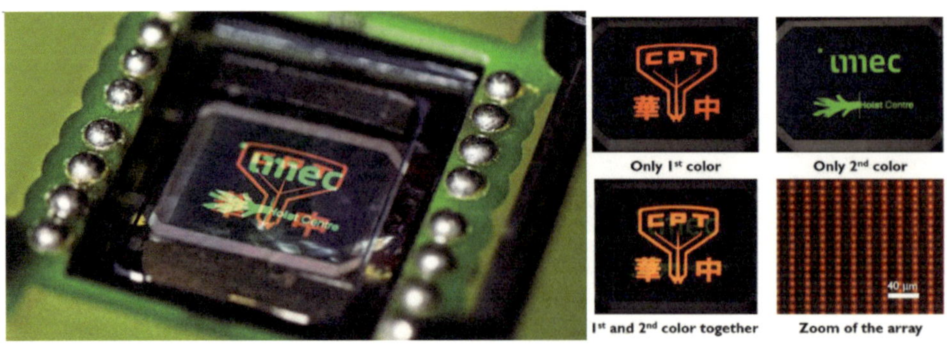

Figure 7.
Passive 1250 ppi patterned OLED display with 1400 × 1400 pixels, 10 μm subpixel pitch, and independent color driving: general view (left) and detailed view for different color drivings (right).

This fabrication process is compatible with both CMOS backplanes and flexible TFT backplanes. The frontplane can thus be implemented in an active matrix display. Of course, photolithography can be used several times to realize more colors for a full-color display.

4. Increased resolution by pixel driving techniques

To increase the display resolution, not only the technology (backplane and frontplane) but also the pixel driving techniques should be optimized. The OLED light output is dependent on the drain current of the driving TFTs of the AMOLED displays. Due to inherent variations in AMOLED displays, some compensation methods to the drain current of the TFTs are required to achieve uniform brightness. This can be implemented through either in-pixel compensation [10] or external compensation [11, 12]. Since in-pixel compensation schemes typically require more transistors inside the pixel, external compensation methods are preferred for high-resolution applications. **Figure 8** shows pixel circuits and a possible layout for in-pixel compensation, using an 8T1C [10] pixel, and external compensation using a 3T2C [11] and, respectively, a 2T1C [12] pixel. For all these layouts, the same design rules were used. It is clear from this figure that a display with external compensation, especially the 2T1C pixel circuit, can achieve a much higher pixel density.

The achievable pixel density depends on both the pixel circuit and the design rules imposed by the technology, such as the critical dimension (CD) of the lithography tool. **Figure 9** compares the achievable resolutions for different CDs for the 8T1C, the 3T2C, and the 2T1C pixel circuit. Although the CD of 1.5 μm, as currently achievable with typical i-line steppers, only yields a maximum pixel density of 565 ppi for the 8T1C pixel circuit, the same CD already yields a significant improvement for the pixel circuits using external compensation, namely, 847 ppi for the 3T2C pixel

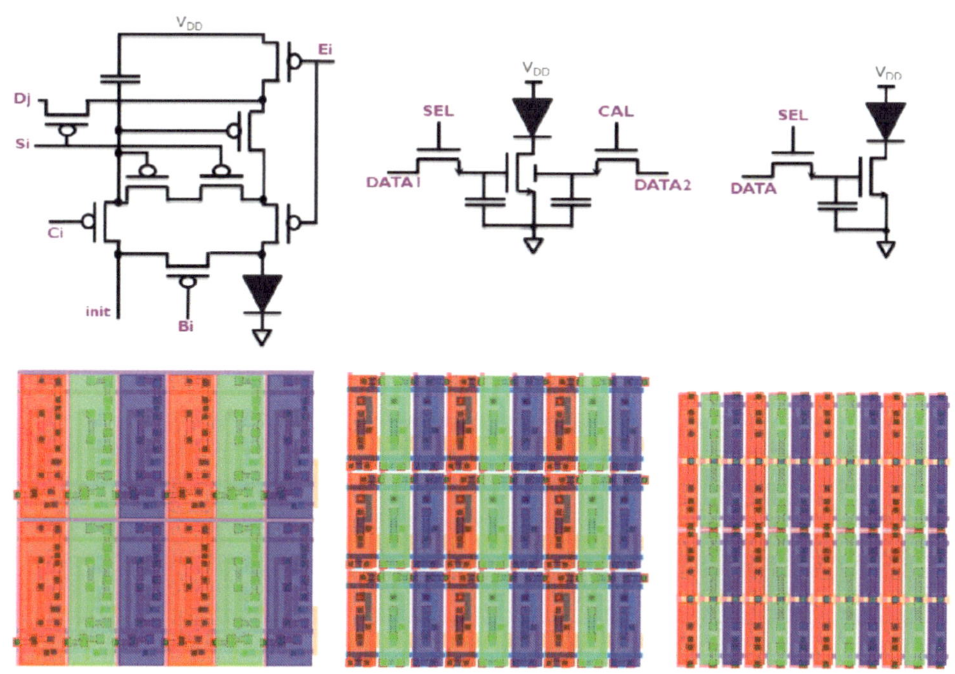

Figure 8.
Pixel circuits and corresponding layouts for (left) 8T1C, (middle) 3T2C, and (right) 2T1C pixels.

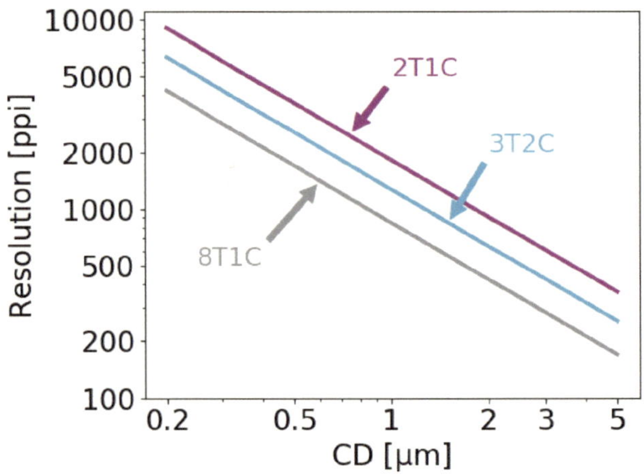

Figure 9.
Pixel resolution vs. critical dimension (CD) for various pixel schemes.

circuit and 1210 ppi for the 2T1C pixel circuit, respectively. Furthermore, improvements in technology allowing smaller CD will even further increase the achievable pixel density, up to 9070 ppi for the 2T1C pixel circuit, when using a CD of 0.2 µm.

The compensation principle for the 3T2C pixel circuit relies on the fact that applying a voltage on the backgate of a transistor will shift the threshold voltage (V_T) of that transistor. By applying the correct compensation voltage to the backgate of the drive transistor of each pixel, all V_T variations can be eliminated, resulting in a more uniform display. This compensation method uses three different modes of operation for the display. The first mode of operation is the calibration mode. In this mode, the correct compensation voltage is determined for each pixel by applying a certain reference voltage to the frontgate and measuring the current through the pixel while varying the backgate. When the measured current matches a predetermined reference current, the voltage on the backgate is the correct compensation voltage, which will be stored both on the capacitor connected to the backgate and in external memory. Once the correct backgate voltage is set for every pixel, the display can be switched to normal operation. In this mode, the display is driven with the normal video data, which is written to the frontgate of each pixel. Since the CAL signal is low in this mode, the charge on the capacitor will remain, and hence the backgate voltage will be the compensated voltage. However, due to leakage, this charge will slowly change over time. Therefore, a third mode of operation is added, namely, the calibration refresh. In this mode, the SEL signal is kept low, but the CAL signal is running through the display, while the compensation data is applied to the data lines. This way the compensation voltage is restored on the backgate, to ensure the V_T uniformity remains over time. This compensation method shows a significant improvement in current variation, as demonstrated in **Figure 10**.

The current through the drive TFT (I_{DS}), and thus through the OLED, when operating in saturation regime can be calculated for a certain data voltage (V_{GS}) by using Eq. (1):

$$I_{DS} = \frac{\mu * C_{ox}}{2} * \frac{W}{L}(V_{GS} - V_T)^2 = \beta * (V_{GS} - V_T)^2 \tag{1}$$

Compensating only for V_T can eliminate variations in current for one gray level; however, if the β-factor is different for each pixel, the current will still vary for different gray values, even after V_T compensation. This is shown in **Figure 11** for the 3T2C pixel. As a consequence, we propose a new compensation method

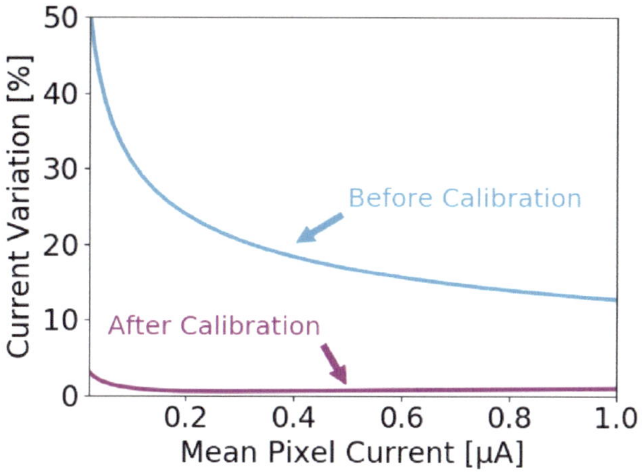

Figure 10.
Current variation of a 3T2C display before and after compensation.

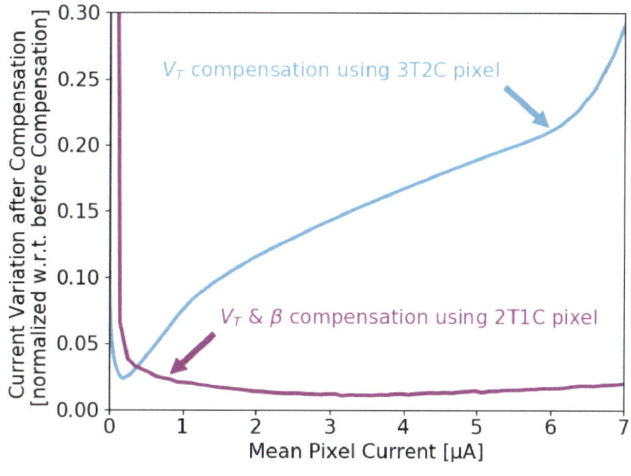

Figure 11.
Current variation of a 3T2C and 2T1C display before and after compensation.

to compensate for both β and V_T variations. Similarly as the previous described compensation method, we will first characterize the current through each pixel for multiple data voltages, whereafter the measurements are fitted to Eq. (1).

For each pixel, the extracted β and V_T values are stored. Based on these values, the V_{GS} voltages can be calculated for each pixel by the driver IC for each desired gray level by using Eq. (2):

$$V_{GS} = \sqrt{\frac{I_{DS}}{\beta}} - V_T \tag{2}$$

This calculation is relative simple and straightforward, as it only requires a multiplication, a subtraction, and a square root calculation, which enables to display real-time video content by using this methodology. **Figure 11** shows the current variation improvements directly obtained from our AMOLED displays, by utilizing the V_T-only compensation method and comparing it to the V_T and β compensation method. As mentioned above, the simple V_T compensation method provides good variation results for a small range, whereas the combined parameter method improves the variation across all desired gray levels.

5. Adding a fourth pixel for finger/palmprint sensing

Fingerprint sensor arrays (Figure 5.1) [13] are becoming a mainstream security mechanism for mobile devices and are today available as autonomous silicon-based component. The integration of the fingerprint sensor array together with AMOLED displays [14–16] would benefit the footprint of the mobile device and the functionality, enabling detection of multiple fingers at once or even a palmprint.

5.1 Side by side/under/over display pixel

Fingerprint sensors combined with AMOLED displays can be realized in three different configurations for the sensor pixels: (1) in the same plane of the display pixels and (2) under and (3) over the display pixels (**Figure 12**). With sensors in the same plane, the display module gains optical sensing capability by incorporating photodetector pixels between OLED pixels. Sensors, under or above the display, require a separate fingerprint module. A fingerprint module under the display would need a semitransparent display and light scattering management. A fingerprint module over the display requires a transparent imager to avoid changes in display emission. In the previous section, we have demonstrated that a higher resolution backplane can be achieved at the same critical dimension by introducing external compensation methods. This combined with the photolitho-based patterning method of OPD and OLED will the crucial enablers for such a configuration as analyzed in Paragraph 3.

5.2 Passive/active sensor pixel

The pixel circuit architecture of the sensor array can be either passive or active [17]. The passive pixel is depicted in **Figure 13** and is comprised by the photoelement, a capacitor, and a select TFT. The main difference of the active pixel is that it requires an extra TFT acting as a local amplifier. For high-resolution applications, the active pixel is not recommended, since its footprint is larger than the footprint of passive pixels.

5.3 Integrated readout circuit

The integration of peripheral readout circuitry in panel side-by-side with the display peripherals is beneficial for resolution, connectivity, and potential lower system cost of the device. In this paragraph TFT-based integrated readout is demonstrated by using our IGZO n-type only TFTs.

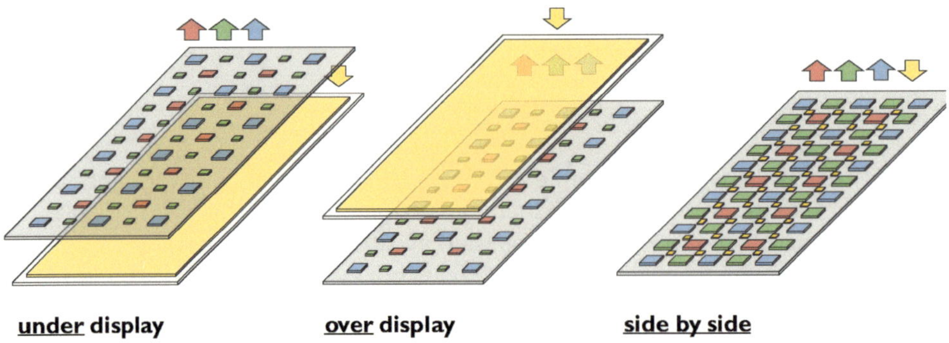

Figure 12.
Fingerprint integration configurations in AMOLED displays.

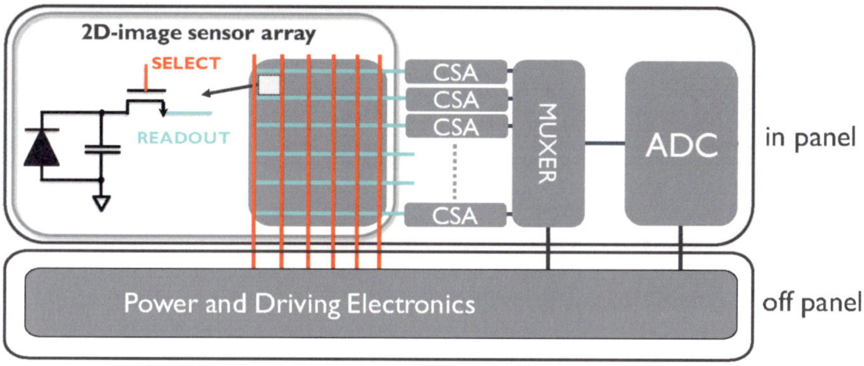

Figure 13.
Block diagram of the proposed in-panel fingerprint sensor array with integrated CSA for each line, multiplexer (MUX), and ADC. Power and driving electronics remain off-panel.

The necessary blocks to implement an in-panel readout system are a charge sense amplifier (CSA), a multiplexer (MUX), and an analog to digital converter (ADC). The CSA is reading out the charge stored in each pixel of the array. The MUX is multiplexing the CSA outputs directly to the ADC, decreasing the number of the required ADC converters. The ADC is converting the analog voltage received from the CSA to a digital code. Various TFT-based analog blocks have been demonstrated in the literature [18–24]. Metal-oxide TFT technologies are preferable due to uniformity over large areas, very low leakage currents, and lower cost over area. In the following section, fast and small footprint ADCs and charge sense amplifiers (CSA) are discussed to meet the specifications of an in-panel fingerprint array.

In **Figure 13**, the high-level block diagram of the in-panel readout system is shown. The five main blocks are detailed: a two-dimensional (2-D) image sensor array, the CSAs connect on each row of the array, a multiplexer connects four or more rows (MUX), and ADC connects to every MUX and off-panel power and driving electronics. Each column of the 2D image sensor array is readout from the corresponding row by a CSA. The pixels are readout subsequently enabled by the "SELECT" signals from the columns and converted to digital code through the MUX and an ADC. The MUX enables a larger footprint for the ADC, up to 4 to 8 times larger compared to a single line (50 μm). Hence, the ADC and MUX needs to be 4 to 8 times faster than the CSA. A 1–2 fps readout of 1M pixel imager (1000 × 1000 pixels) sets a readout speed of 1–2 kS/s per line for each CSA. This translates to 4–8 kS/s for the ADC if a 4:1 MUX is used or 8–16 kS/s for a 8:1 MUX. The slower ADC configuration sets the specification limitation to the width of the ADC to 200 μm, whereby the faster allows a width of 400 μm.

5.4 Dual-gate metal-oxide technology

Figure 14 (a) depicts the cross section of the dual-gate self-aligned metal-oxide (MO) technology on a 15-μm-thick polyimide film [25, 26]. The metal-oxide (IGZO or ITZO) TFTs are fabricated with two metal gates (M0, M1) and source-drain metal contacts (M2). An additional metal layer (M3), not shown in the cross section, is beneficial for footprint but also for performance and noise. The CSA experimental results shown in the following sections are designed with an extra metal layer, also used as anode layer.

Figure 14(b) shows the distribution of the extracted on-current (I_{ON}) from the measured transfer characteristics of 480/20 (μm/μm) IGZO (red) and ITZO (orange) dual-gate TFTs. The IGZO TFTs exhibits a median I_{ON} of 54.4 μA, whereas

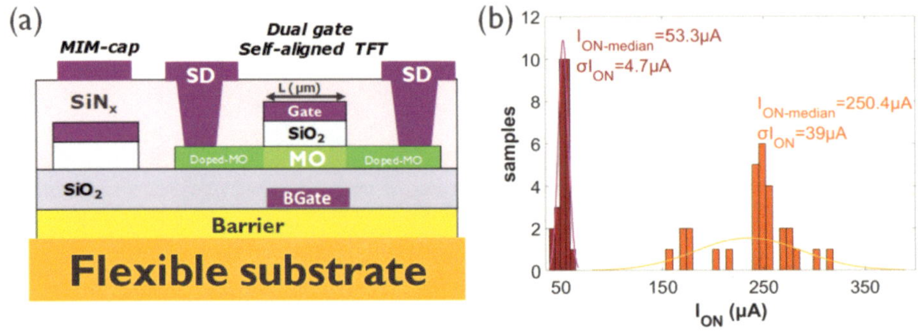

Figure 14.
(a) Cross section of dual-gate self-aligned metal-oxide technology on flexible polyimide substrate and (b) extracted on-current (I_{ON}) from experimental data of 480/20 (μm/μm) dual-gate self-aligned ITZO (red) and IGZO (orange) TFTs.

the ITZO dual-gate TFT exhibits a median of 250.4 μA. Although the I_{ON} of the ITZO TFT is 5 times larger, the normalized spread of I_{ON} to the median over the wafer is double (15.4% for ITZO and 8.8% for IGZO). Threshold voltage is also extracted from the same measurements, yielding 1.16 V (and σV_t = 242 mV) for ITZO and 1.77 V (and σV_t = 94 mV) for IGZO TFTs. The 480/20 TFT is the largest footprint TFT used in the implemented designs. In **Figure 15**, microphotos of the (a) ADC and (b) the L = 3 μm CSA are shown.

5.5 Charge sense amplifier

The schematic of the CSA is shown in the **Figure 16(a)**. The schematic of the operational amplifier (OPAMP) used in the CSA is shown in **Figure 16(b)**. The OPAMP comprises a differential pair and a load that is driven by a two-stage buffer, initiated by a start-up circuit. The driver consists of two diode-connected load inverters as input and output stages. The output stage is driven by a start-up circuit to initialize the operation and provide ~1 loop gain positive feedback bias to the n-type load of the differential amplifier such that gate voltage follows source voltage. A buffer is also included at the output nodes A and B to increase the speed of the amplifier.

The experimental results of open-loop experiments of the OPAMP are shown in **Figure 17** for dual-gate self-aligned IGZO TFTs of minimum channel length of L = 5 μm. The maximum gain for the 5 μm design is 43.2 dB with a phase margin of 52°. Both parameters are critical for stable closed-loop operation of an OPAMP. The

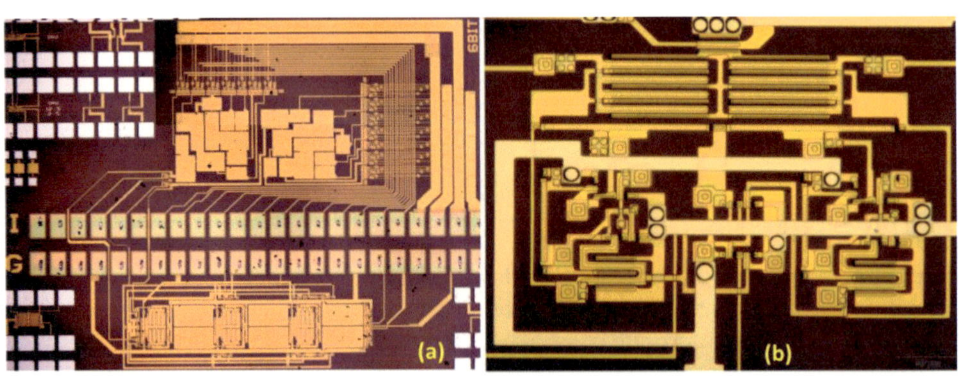

Figure 15.
Microphotos of the (a) ADC and (b) L = 3 μm CSA on flexible substrate.

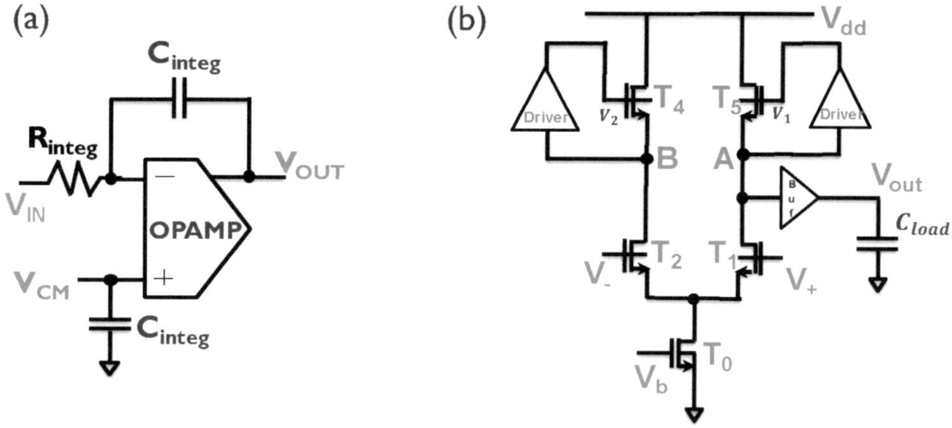

Figure 16.
Schematic of (a) charge sense amplifier (CSA) and (b) the OPAMP schematic using dual-gate self-aligned TFTs used in the CSA in (a).

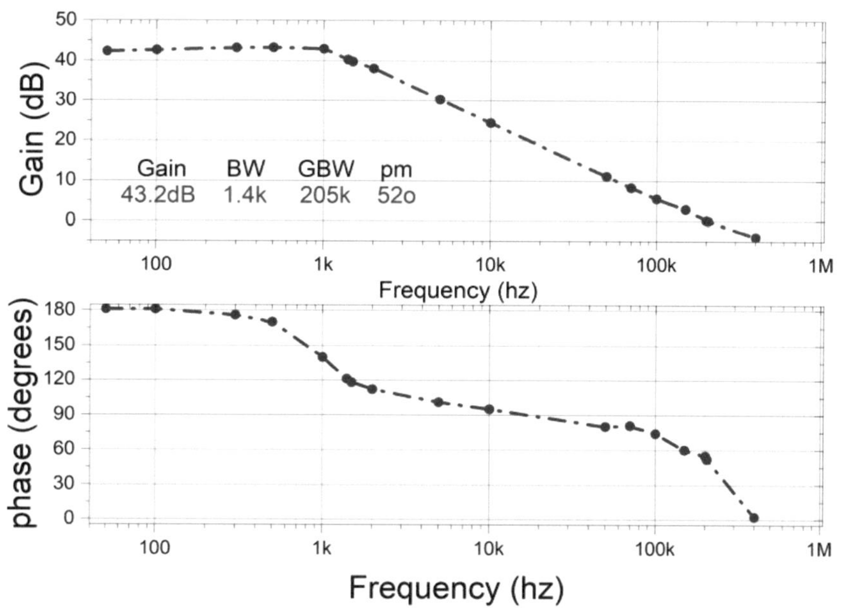

Figure 17.
Experimental bode plots of the OPAMP using dual-gate self-aligned IGZO technology for L = 5 μm.

obtained bandwidth (BW) is 1.4 kHz, and gain-bandwidth reaches 205 kHz. These specs are compared to other publications in the state-of-the-art **Table 1** using TFTs. The footprint of the CSA is 0.28 mm^2 with the capacitors. The footprint of the CSA can be decreased to 0.07 mm^2 if 3 μm design is implemented. These footprints will result in a bezel width of 5.5 mm (L = 5 μm) and 1.3 mm (L = 3 μm) [27] for a 50 μm pixel size.

5.6 Analog to digital converter

The successive approximation C-2C architecture is selected as ADC architecture, due to the low power dissipation and the 0.1% uniformity of metal-insulator-metal capacitors across large-area thin-film wafers. The schematic of the ADC is shown in **Figure 18**. The comparator is the most critical building block of the ADC for speed

Circuit	This work	2018 [11]	2012 [8]	2013 [10]	2014 [9]
Supply (V)	**15**	15	10.5	5	50
Gain (dB)	**43**	32	~21	18.7	19.2
GBW (kHz)	**205**	140	2	472	100*
pm (°)	**52**	53	50*	neg	46
Area (mm^2)	**0.28**	0.3	*Discrete*	2.08	—
Substrate	**PI**	PI		PI	Glass
L (μm)	5	5	10	6	10

***Estimated.**

Table 1.
In-Ga-Zn-O-TFT differential amplifier comparison.

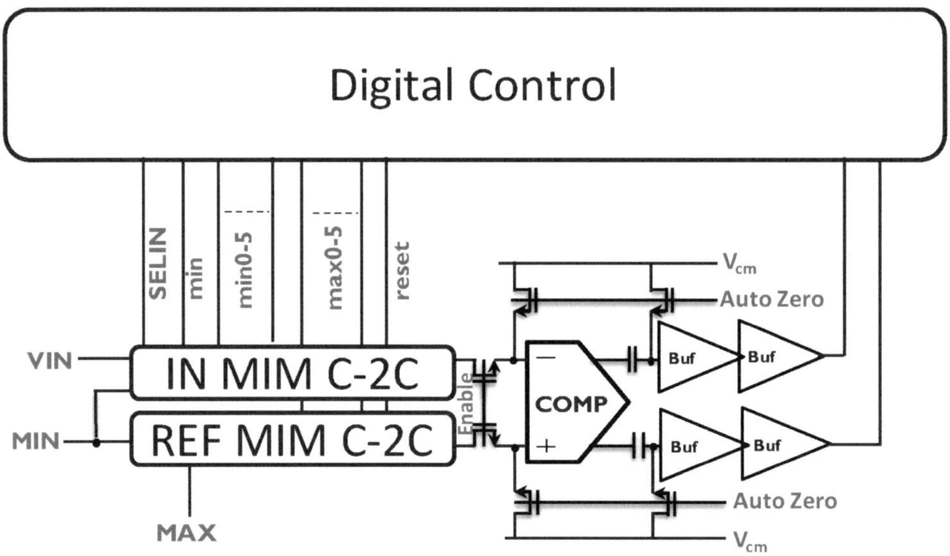

Figure 18.
The implemented SADG TFT ADC block diagram driven with offset compensation.

and accuracy. To improve the accuracy of the comparator, open-loop offset cancelation [28] is used. The response of a L = 5 μm dual-gate self-aligned TFT-based comparator at 10 kHz for two inputs ΔV_{in} = 0 and 10 mV is shown in **Figure 19**. The offset of the comparator is minimized to less than 10 mV using open-loop cancelation, even though the V_t variation of the TFTs is one order larger.

The experimental results of the reconstructed samples from a sinusoidal analog wave applied at the input of the IGZO ADC are shown in **Figure 20(a)**. The IGZO ADC achieves 6-bit resolution at a sampling speed of 133 S/s at 15 V power supply using a L = 20 μm comparator. The clock speed of the IGZO ADC is at 2 kHz matching the bandwidth of the comparator, and 15 clocks are required to complete the conversion. Two options are available to increase the sampling speed of the ADC as set by the specification for 1–2 fps fingerprint readout: channel length downscaling and/or change of TFT technology or change of ADC architecture to flash but then for 6-bit the area might increase dramatically due to the multiple resistors needed and 64 comparators.

The minimum length of the TFTs of the measured ADC is L = 20 μm. The length of the TFT of the comparator of the ADC defines the speed of the circuit.

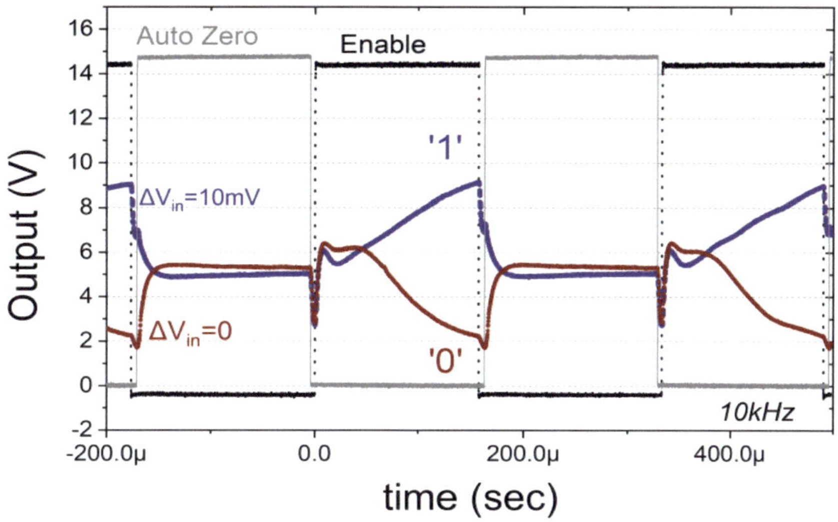

Figure 19.
Measured comparator output with auto-zero offset cancelation for designs using TFTs of minimum L = 5 μm at 10 kHz.

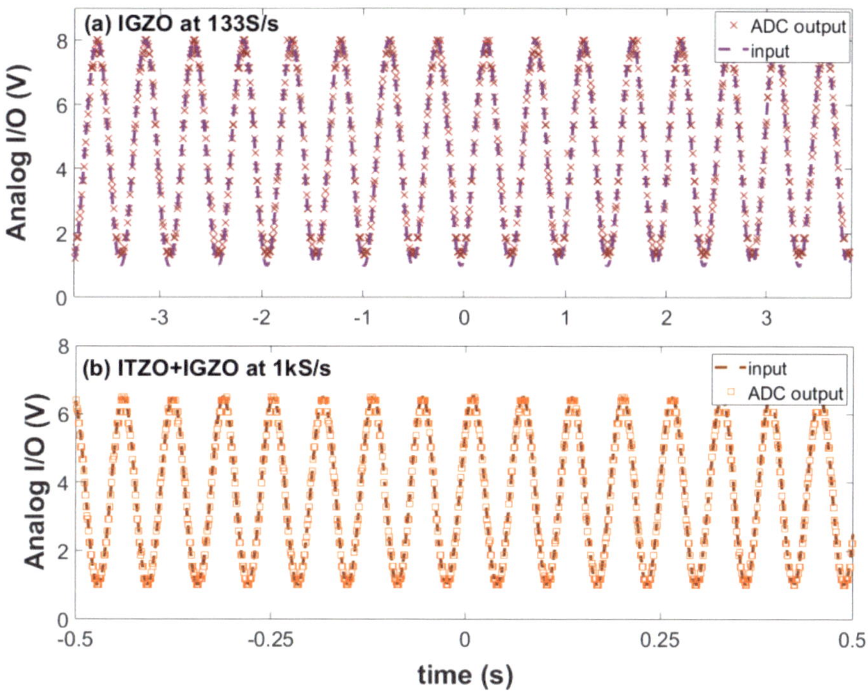

Figure 20.
(a) The applied analog input of 2.061 Hz sinewave signal to the ADC and the reconstructed output points from the digital output of the ADC at a clock frequency of 2 kHz for IGZO and (b) the applied analog input of 15.625 Hz sinewave to the 15 kHz clocked ITZO ADC and the reconstructed output.

In **Figure 19** the response of the L = 5 μm comparator is shown validating 10 kHz operation, 5 times faster compared to the 20 μm comparator used in the ADC. This indicates that 5 μm comparator can increase the sampling speed by 5 times.

Another option to enable a larger increase of the ADC speed is to introduce ITZO TFT for the ADC. ITZO TFT exhibits 5 times larger I_{ON} which leads to faster responses [**Figure 20(b)**]. The ITZO ADC achieves a similar bit resolution of the applied sinusoidal wave at 1 kS/s sampling speed and at 10 V supply voltage. Combining both ITZO and downscaling of the length of the TFTs of the comparator will lead to sampling speeds above 4 kS/s as required for the in-panel readout system of 1 fps. The power dissipation of the ITZO ADC is at 550 μW at 10 V power supply and at 110 μW at 15 V power supply for the IGZO implementation.

In conclusion, TFT-based in-panel analog circuits for 1 fps readout of a fingerprint or palmprints array is presented in this paragraph. The circuit blocks can be integrated side-by-side to the flat panel display with integrated sensor array. Two analog blocks are discussed, being an analog to digital converter and charge sense amplifier. ADCs and CSA downscaling to 5 μm including a 500% I_{ON} boost of the ITZO TFTs enable the 4 kS/s operation specifications for the in-panel readout circuits. The use of complementary technologies such as LTPS or LTPO would result an increased performance for important parameters such as the gain-bandwidth and the resolution and speed of the ADC. This will enable a better optimized system for this application.

6. Conclusions

In this book chapter, we have discussed a roadmap to include additional functionalities on displays by adding a fourth pixel to the display. The purpose of the fourth pixel in this work was a photodetector pixel to realize an in-panel fingerprint or palmprinting function. Several options have been discussed to enable this roadmap. At first, we have elaborated several techniques to realize higher-resolution frontplane (OLED) and TFT backplanes. For the frontplanes, photolitho patterning of the OLED was introduced as a disruptive technology impacting significantly the frontplane resolution. External compensation techniques for the backplane when driving OLEDs are proposed in this book chapter, resulting in a uniform and higher-resolution display by using only 2T1C schemes. As photodetector pixel, we proposed a passive 1 T pixel, with included peripheral circuits to enable a lower system cost.

Author details

Nikolaos Papadopoulos[1*], Pawel Malinowski[1], Lynn Verschueren[1,2], Tung Huei Ke[1], Auke Jisk Kronemeijer[3], Jan Genoe[1,2], Wim Dehaene[1,2] and Kris Myny[1]

1 Imec, Heverlee, Belgium

2 ESAT Department, KU Leuven, Leuven, Belgium

3 TNO, Holst Center, Eindhoven, The Netherlands

*Address all correspondence to: nikolas.papadopoulos@imec.be

References

[1] Balram N, Tošić I. Light-field imaging and display systems. Information Display. 2016;**32**(4):6-13

[2] Reichelt S, Häussler R, Fütterer G, Leister N. Depth cues in human visual perception and their realization in 3D displays. In: Proceedings of SPIE 7690, Three-Dimensional Imaging, Visualization, and Display 2010 and Display Technologies and Applications for Defense, Security, and Avionics IV, 76900B, 14 May 2010. 2010

[3] Tsushima Y. Super Hi-Vision (8K) Produces Stronger Depth Sensation than 4K and Hi-Vision (2K). Japan: Niigata; 2014

[4] Kawashima S, Shishido H, Oshita S, Yoshitomi S, Yamashita A, Ishitani T, et al. 18-5: A 1058-ppi 4K ultrahigh-resolution and high aperture LCD with transparent pixels using OS/OC technology. SID Symposium Digest of Technical Papers. 2017;**48**(1):242-245

[5] Ghosh A, Donoghue EP, Khayrullin I, Ali T, Wacyk L, Tice K, et al. 18-1: Invited paper: Ultra-high-brightness 2K x 2K full-color OLED microdisplay using direct patterning of OLED emitters. SID Symposium Digest of Technical Papers. 2017;**48**(1):226-229

[6] Zakhidov A, Lee J-K, DeFranco J, Hang Fong H, Taylor PG, Chatzichristidi M, et al. Orthogonal processing: A new strategy for organic electronics. Chemical Science. 2011;**2**(6):1178-1182

[7] Malinowski PE, Ke Atsushi T-H, Vicca NP, Kronemeijer AJ, Laurens MAJ, van der Steen S, et al. 44-1: Invited Paper: Photolithography as Enabler of AMOLED Displays beyond 1000 ppi - Malinowski [Internet]. SID Symposium Digest of Technical Papers/Wiley Online Library. 2017. Available from: https://onlinelibrary.wiley.com/doi/abs/10.1002/sdtp.11715 [cited 14 January 2020]

[8] Ke TH, Malinowski P, Nakamura A, Wuyts M, Gu D, Lee J-H, et al. Effect of integrated protection layer on photolithographic patterned organic light emitting diodes. In: 23rd International Display Workshops—IDW [Internet]. 2016. pp. 1187-1188. Available from: https://lirias.kuleuven.be/1733178 [cited 14 January 2020]

[9] Malinowski PE, Ke T-H, Nakamura A, Liu Y-H, Vander Velpen D, Vandenplas E, et al. High resolution photolithography for direct view active matrix organic light-emitting diode augmented reality displays: Organic light-emitting diode lithography as enabler for hi-res displays. ournal of the Society for Information Display. 2018;**26**(3):128-136

[10] Kim C. Pixel Circuit and Organic Light-Emitting Diode Display Including the Same [Internet]. 20170148383. USA: United States Patent and Trademark Office; 2017. Available from: http://www.freepatentsonline.com/y2017/0148383.html [cited 14 January 2020]

[11] Verschueren L, Ameys M, De Roose F, Steudel S, van der Steen J-L, Gelinck G, et al. 35-2: 40x current variation reduction enabled by an external VT-compensation scheme for AMOLED displays using a 3T2C-pixel circuit with dual-gate TFTs. SID Symposium Digest of Technical Papers. 2018;**49**(1):437-440

[12] Verschueren L, Ameys M, Velazquez Lopez M, Smout S, Ke TH, Vandenplas E, et al. 38-4: A 2T1C AMOLED display with external compensation reducing on-panel current variations to 0.079 percent. In: SID Symposium Digest of Technical Papers. 2020

[13] Akkerman H, Peeters B, van Breemen A, Shanmugam S, Tordera D,

van der Steen J-L, et al. 39-3: Printed organic photodetector arrays and their use in palmprint scanners. SID Symposium Digest of Technical Papers. 2018;**49**(1):494-497

[14] Qin Y, Wang H, Liu Y. P-215: Organic-inorganic hybrid thin-film photo-detector for fingerprint recognition. SID Symposium Digest of Technical Papers. 2018;**49**(1):1604-1606

[15] Zhou X, Zhang M, Xu Y, Zhou W, Wang K, Nathan A, et al. 39-2: Highly sensitive a-Si:H PIN photodiode gated LTPS TFT for optical In-display fingerprint identification. SID Symposium Digest of Technical Papers. 2018;**49**(1):490-493

[16] Cho A-T, Liu Z, Liu K, Yang F, Mo Q, Hsu J, et al. 76-4: Late-news paper: Flat panel fingerprint/touch-input image sensor using a-Si TFT photo-transistor and four-mask process architecture technology. SID Symposium Digest of Technical Papers. 2018;**49**(1):1024-1027

[17] Karim KS, Nathan A, Rowlands JA. Amorphous silicon active pixel sensor readout circuit for digital imaging. IEEE Transactions on Electron Devices. Jan 2003;**50**(1):200-208

[18] Lin W-M, Liu S-I, Lin C-F. A CBSC Second-Order Sigma-Delta Modulator in 3um LTPS-TFT Technology. Taipei, Taiwan: IEEE; 2009. pp. 133-136

[19] Papadopoulos NP, De Roose F, van der Steen J-LPJ, Smits ECP, Ameys M, Dehaene W, et al. Toward temperature tracking with unipolar metal-oxide thin-film SAR C-2C ADC on plastic. IEEE Journal of Solid-State Circuits. 2018;**53**(8):2263-2272

[20] Tai Y-H, Chiu H-L, Chou L-S, Chang C-H. Boosted gain of the differential amplifier using the second gate of the dual-gate a-IGZO TFTs. IEEE Electron Device Letters. 2012;**33**(12):1729-1731

[21] Zysset C, Munzenrieder N, Petti L, Buthe L, Salvatore GA, Troster G. IGZO TFT-based all-enhancement operational amplifier bent to a radius of 5 mm. IEEE Electron Device Letters. 2013;**34**(11):1394-1396

[22] Garripoli C, van der Steen J-LPJ, Smits E, Gelinck GH, Van Roermund AHM, Cantatore E. 15.3 an a-IGZO asynchronous delta-sigma modulator on foil achieving up to 43dB SNR and 40dB SNDR in 300Hz bandwidth. In: 2017 IEEE International Solid-State Circuits Conference (ISSCC) [Internet]. San Francisco, CA, USA: IEEE; 2017. pp. 260-261. Available from: http://ieeexplore.ieee.org/document/7870360/ [cited 14 January 2020]

[23] Papadopoulos N, Steudel S, Roose FD, Eigabry DM, Kronemeijer AJ, Genoe J, et al. In-panel 31.17dB 140kHz 87μW unipolar dual-gate In-Ga-Zn-O charge-sense amplifier for 500dpi sensor Array on flexible displays. In: ESSCIRC 2018—IEEE 44th European Solid State Circuits Conference (ESSCIRC). 2018. pp. 194-197

[24] Kim K, Choi K-Y, Choi D-H, Lee H. P-51: A-IGZO TFT based operational amplifier and comparator circuits for the adaptive DC-DC converter. SID Symposium Digest of Technical Papers. 2014;**45**(1):1164-1167

[25] Upadhyay R, Steudel S, Hung M-P, Mandal AK, Catthoor F, Nag M. Self-aligned amorphous indium-tin-zinc-oxide thin film transistors on polyimide foil. ECS Journal of Solid State Science and Technology. 2018;7(4):P185-P191

[26] Kronemeijer AJ, Akkerman H, Steen J-L van der, Steudel S, Pendyala R, Panditha P, et al. P-127: Dual-gate self-aligned IGZO TFTs monolithically integrated with high-temperature bottom moisture barrier for flexible AMOLED. SID Symposium Digest of Technical Papers. 2018;**49**(1):1577-1580

[27] Papadopoulos N, Steudel S, Smout S, Willegems M, Nag M, Ameys M, et al. 9-1: Invited paper: Metal-oxide readout electronics based on indium-gallium-zinc-oxide and indium-tin-zinc-oxide for in-panel fingerprint detection application. SID Symposium Digest of Technical Papers. 2019;**50**(1):95-98

[28] Enz CC, Temes GC. Circuit techniques for reducing the effects of op-amp imperfections: Autozeroing, correlated double sampling, and chopper stabilization. Proceedings of the IEEE. 1996;**84**(11):1584-1614

Chapter 6

Vertical-Type Organic Light-Emitting Transistors with High Effective Aperture Ratios

Byoungchoo Park, Won Seok Lee, Seo Yeong Na, Jaewoo Park and In-Gon Bae

Abstract

The inherent complexity of the structures of active-matrix (AM) organic light-emitting diode (OLED) displays severely limits not only their size but also device performance. Surface-emitting organic light-emitting transistors (OLETs) may offer an attractive alternative to AM displays. We report some characteristics of vertical-type OLETs (VOLETs) composed of a source electrode of low-dimensional materials and an emissive channel layer. With a functionalized graphene source, it is shown that the full-surface electroluminescent emission of a VOLET can be effectively controlled by the gate voltage with a high luminance on/off ratio (10^4). The current efficiency and effective aperture ratios were observed to be more than 150% of those of a control OLED, even at high luminances exceeding 500 cd m^{-2}. Moreover, high device performance of micro-VOLET pixels has been also successfully demonstrated using inkjet-patterned emissive channel layers. These significant improvements in the device performance were attributed to the effective gate-voltage-induced modulation of the hole tunneling injection at the source electrode.

Keywords: organic light-emitting diode (OLED), organic light-emitting transistor (OLET), vertical-type OLETs, graphene, on/off ratio, aperture ratio, inkjet printing, tunneling injection

1. Introduction

In recent years, researchers of state-of-the-art electronics utilizing organic semiconducting materials have succeeded in advancing various devices, such as organic light-emitting diodes (OLEDs), photovoltaic cells, organic thin-film transistors (OTFTs), and sensors, among others [1–10]. Among these, intensive efforts in OLEDs have led to high brightness, efficiency, and full-color electroluminescent (EL) emissions for various light-emitting optoelectronic devices [7–10]. The advantages of such OLEDs over conventional liquid crystal displays (LCDs) are well known, especially for high-quality displays in terms of their viewing angle, response time, thickness, and contrast ratio [11]. For instance, small OLED displays are constructed on an array of thin-film transistor (TFT) switches, allowing precise control of the states of the pixels [12–14]. In such active-matrix OLEDs (AM-OLEDs), the OLED is driven in the current mode; thus, at least two TFTs, in this case a switching TFT to select a pixel and a driving TFT to operate the OLED, are required, as shown in **Figure 1(a)** [12, 13]. Perhaps unexpectedly, however, the complexity of such pixel circuit designs with their sophisticated procedures has led to a significantly limited light-emitting area and aperture ratio (the light-emitting area as a fraction of the total area of the device, typical aperture ratios: 25–34%) [13–15], introducing severe problems associated with limited device performance and limited display sizes for AM-OLEDs. Besides these issues, fundamental factors related to the architecture of the OLED itself, such as exciton quenching and photon loss, also still limit the efficiency and brightness of these devices.

To overcome some of the limitations of (AM-)OLEDs, research on different structures and materials is currently yielding new developments [15–30]. Among these, organic light-emitting transistors (OLETs), such as static-induction-transistor OLETs (SIT-OLETs) [17, 18], metal-insulator-semiconductor OLETs (MIS-OLETs) [19], lateral-type OLETs [20–29], and vertical-type OLETs (VOLETs) [30], have been devised by integrating the capability of the OLED to generate EL light with the switching functionality of a field-effect transistor (FET) into a single device structure. In these OLETs, the current that flows through emissive semiconductor channel layers can be controlled by the gate voltage, which can also change the EL emission brightness state from the dark off- to the bright on-state. The on-state implies that holes and electrons injected into the channel layer form excitons that recombine radiatively to generate EL light [17–30]. These OLETs are of key interest; not only do they provide a novel device architecture to investigate fundamental optoelectronic properties related to charge carrier injection, transport, and radiative exciton recombination processes in organic semiconducting materials, at the same time OLETs can also be used to develop highly integrated organic optoelectronic devices such as highly bright and efficient light sources, optical communication systems, and electrically driven organic lasers [21–30].

In principle, the luminance from OLETs can be modulated by the gate voltage without any additional driving devices; thus, displays using OLETs have the advantage of greatly reducing both the number of TFTs and the circuit complexity (**Figure 1(b)**), thereby providing an effective means of increasing the aperture ratio [29]. Hence, OLETs could be a key part of the development of next-generation AM display technology [29]. Indeed, a proof-of-principle device was recently developed using carbon nanotubes (CNTs, **Figure 2**), delivering a CNT-based vertical-type OLET (CNT-VOLET) [31–34]. In the CNT-VOLET, a dilute network of CNTs is used as a source electrode, leading to several improvements, such as a high on/off ratio, attributed to the gate-bias-induced modulation of the lateral (or horizontal) Schottky barrier height [31, 32]. Nevertheless, the improvement of the effective aperture ratio (AR_{eff}, the percentage ratio of the current efficiency of a surface emitting OLET to that of a control ITO-base OLED at a given luminance) in the

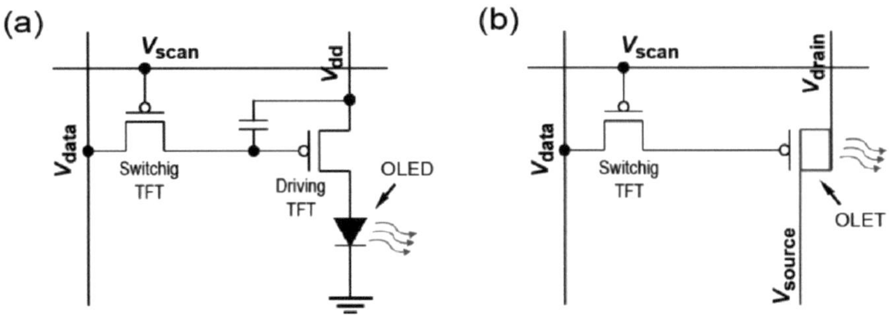

Figure 1.
(a) A conventional two TFTs + one capacitor AM-OLED pixel circuit diagram with a switching TFT and driving TFT. (b) A simple AM-OLET pixel circuit of an integrated OLET and a switching TFT.

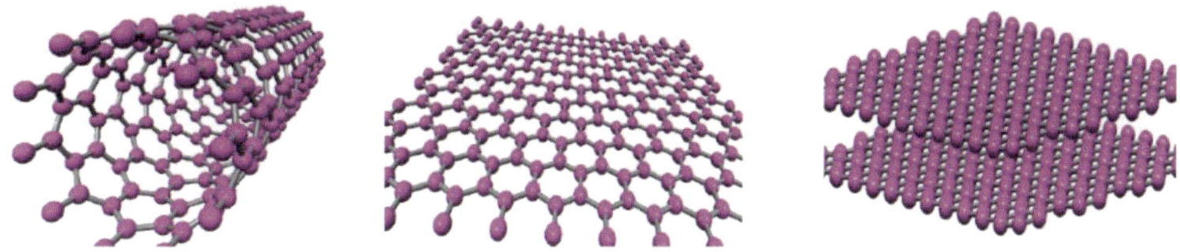

Figure 2.
The structure of carbon nanotube (CNT), single-layer graphene (SLG), and bilayer graphene (BLG).

CNT-VOLET is still limited to less than those of control OLEDs, and its parasitic power consumption (*PPC*, percentage of power dissipated across the driving transistor element of the device not contributing to light generation) requires further reductions [32]. Moreover, this attempt resulted in a complicated source structure, and the production of porous CNT network sources with smooth and homogeneous surfaces was problematic due to the aggregation of CNTs [35]; significant obstacles thus remain with regard to the limited reproducibility of these devices. Thus, the goal of high-performance and reliable OLETs with high AR_{eff} and low *PPC* values remains a considerable challenge.

In this chapter, for the VOLET, the use of a nonporous, homogeneous, smooth, and easily processable graphene layer is described as the source contact, together with an emissive channel layer. Here, the graphene is a two-dimensional (2D) material in the form of a single atomic layer of carbon with a hexagonal lattice structure bonded in the sp^2 configuration (**Figure 2**) [36–38]. Despite the similar low dimensionality of graphene to CNTs [36, 37], the optoelectric properties of a VOLET based on graphene have not yet been fully characterized. Herein, the fabrication and characterization are described for a simple VOLET with a single-layer graphene (SLG) source contact (Gr-VOLET), capable of efficiently modulating device performance levels with high luminance on/off ratios (~10^4) upon the application of a gate voltage. The Gr-VOLETs with doped SLG sources with $FeCl_3$ are demonstrated to exhibit greatly improved device performance, especially in their enhanced current efficiency and AR_{eff} values of more than 150% of those of a control OLED, even at high EL luminance levels exceeding 500 cd m^{-2}. These figures are among the highest ever published for OLETs, and their low *PPC*s make them all the more attractive. Moreover, such high device performance has also been successfully confirmed even for micro-VOLET pixels fabricated by an inkjet-patterning technique [39].

2. 2D material electrode: doped CVD graphene

2.1 Preparation of SLG source electrodes

Preparation of substrates: The transparent VOLET substrate used here was prepared with a pre-patterned bottom gate electrode, consisting of a transparent ITO layer (80-nm-thick, 30 ohm $square^{-1}$ sheet resistance) on a glass substrate, and a sputter-deposited aluminum oxide (Al_2O_3) top layer (400 nm) as a gate dielectric over the ITO gate (glass/ITO/Al_2O_3). The VOLET substrate used was then cleaned with alcohol, followed by a UV treatment (5 min), prior to the fabrication of the devices.

Synthesis and transfer of SLG: The procedure used for transferring the chemical-vapor-deposition (CVD)-grown graphene onto a target substrate [40–45], in this case an FET substrate, a VOLET substrate, or a glass substrate, is described below. The first step involves the CVD growth of graphene on a copper (Cu) foil [42–45]. A clean Cu foil was placed in a quartz tube vacuum chamber and then the temperature of the chamber was increased to 1000°C under Ar (10 sccm). For the growth of graphene, a mixed gas of methane (CH_4, 30 sccm) and hydrogen (H_2, 10 sccm) was used at approximately 2.7×10^{-2} Pa. The next step involved spin-coating a solution of poly-methyl methacrylate (PMMA) at 3000 rpm for 60 s onto the graphene on the Cu foil [42–45]. The graphene on the back side of the Cu foil was removed by atmospheric-pressure oxygen plasma. Next, a PMMA-coated Cu/Gr (Cu/Gr/PMMA) block (width: 4 mm, length: 20 mm) was floated on an aqueous $FeCl_3$ etching solution used to etch the Cu foil entirely, at 50°C for 20 min [40]. Next, the Gr/PMMA block was rinsed with deionized (DI) water two to five times (10 min) and transferred onto a target substrate, after which the graphene-transferred substrate was dried under reduced pressure (~1 Pa) for 1 h and left in air for 24 h. The PMMA layer was then removed by dissolving the PMMA layer in chloroform (~60 min), monochlorobenzene (~30 min), and chloroform again (~30 min), to obtain a SLG source electrode. The optical characteristics of the SLG-transferred VOLET substrate were monitored using a UV–visible spectroscopy system. The average optical transmittance (~92%) of a SLG source on a VOLET substrate in the visible range (400–800 nm) was found to be similar to that (~92%) of a conventional ITO-coated glass substrate for OLEDs.

2.2 Characterization of SLGs

In this study, a transferred SLG was investigated as a source contact, where the $FeCl_3$ doping is processed spontaneously during the graphene transfer process [40]. The basic properties of the three SLG sources are shown in **Figures 3** and **4** and are summarized in **Table 1**.

In order to identify the SLG used, the surface composition of the SLG on the SiO_2/Si substrates was analyzed by X-ray photoelectron spectroscopy (XPS). **Figure 3(a)** presents the wide-scan XPS spectra, showing strong photoelectron lines at binding energies of ~104, ~285, and ~531 eV, which are attributed to Si2p, C1s and O1s, respectively. Note that there is no Cu peak in the range of 932–935 eV (Cu2p and Cu^{2+}), implying the complete etching of the Cu foil. In addition, the XPS spectra also revealed small but measurable amounts of Cl and Fe. These are likely residues of the etchant ($FeCl_3$) used during the etching process. When such $FeCl_3$ residues adsorb onto the SLG, the transfer of electrons to the Cl from the SLG (chlorination) [46] induces unintentional p-doping of the SLG.

Figure 3(b) shows the surface morphology of the SLG on the VOLET substrate as measured by a noncontact atomic force microscope (AFM). As indicated by the AFM morphology, the SLG samples exhibit a fairly smooth surface; the SLG presented an AFM morphology that was nearly identical at different positions on the investigated SLG samples with low RMS roughness levels of 1.4–2.0 nm.

The surface-contact potential difference (V_{CPD}) of the SLG on the VOLET substrate was also monitored using a simultaneous Kelvin prove force microscope (KPFM) by applying AC voltage (1 V) with a frequency of 18 kHz to a Pt/Ir-coated silicon KPFM cantilever. In order to calibrate the work function of the sample, V_{CPD} of highly oriented pyrolytic graphite (HOPG) was used as a reference V_{CPD}. The work function of the SLG (W_{SLG}) was obtained by a comparison of V_{CPD}s for the SLG and the HOPG,

$$W_{SLG} = W_{HOPG} + [V_{CPD}(HOPG) - V_{CPD}(SLG)], \quad (1)$$

where W_{HOPG} is the work function of the HOPG (~4.6 eV) [47]. **Figure 3(c)** shows the distributions of the work functions of the SLG as measured by the KPFM. The estimated average work function of the $FeCl_3$-doped SLG on the VOLET substrate is approximately 5.21 ± 0.07 eV, which is higher than the intrinsic work function (4.5–4.8 eV) of undoped monolayer graphenes [47, 48], mainly due to the $FeCl_3$ doping [49].

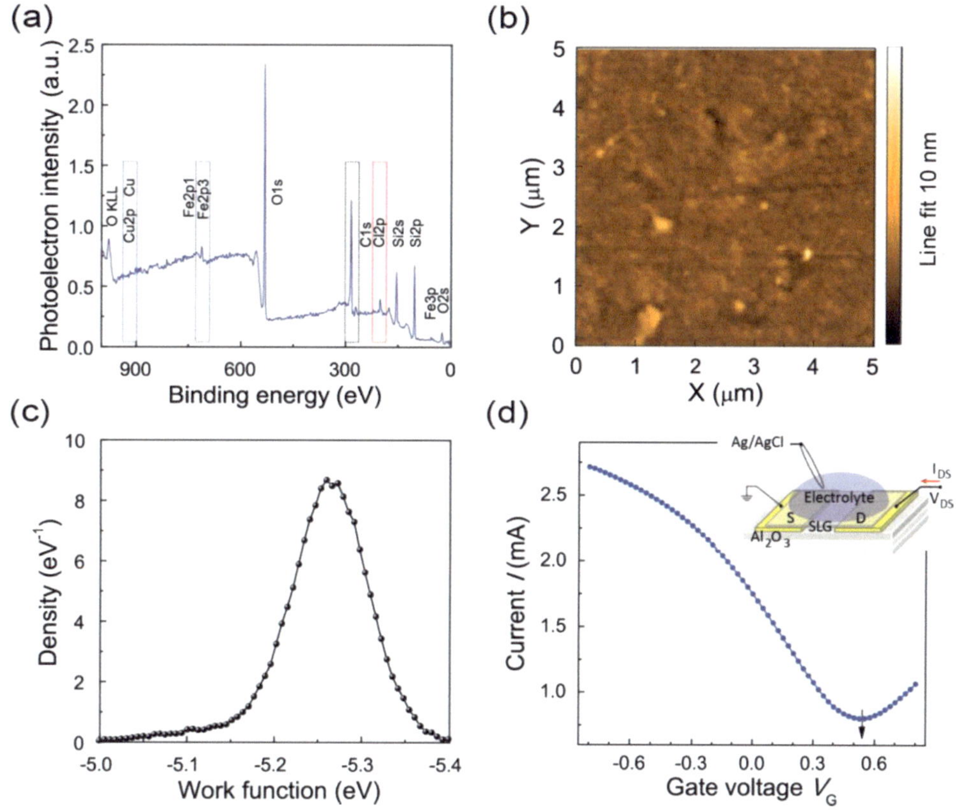

Figure 3.
(a) Wide-scan XPS survey spectra of the studied SLG on a SiO_2/Si substrate. AFM topographic image (5 μm × 5 μm) (b) and corresponding work function distribution (c) of the SLG on the VOLET substrate as measured by KPFM. (d) Transport characteristics of the SLG from liquid-gated lateral Gr-FETs at V_{DS} = −100 mV. The insets show the structure of the liquid-gated lateral Gr-FET with an Ag/AgCl reference electrode in a nonaqueous electrolyte containing ACN and 100 mM of $TBAPF_6$.

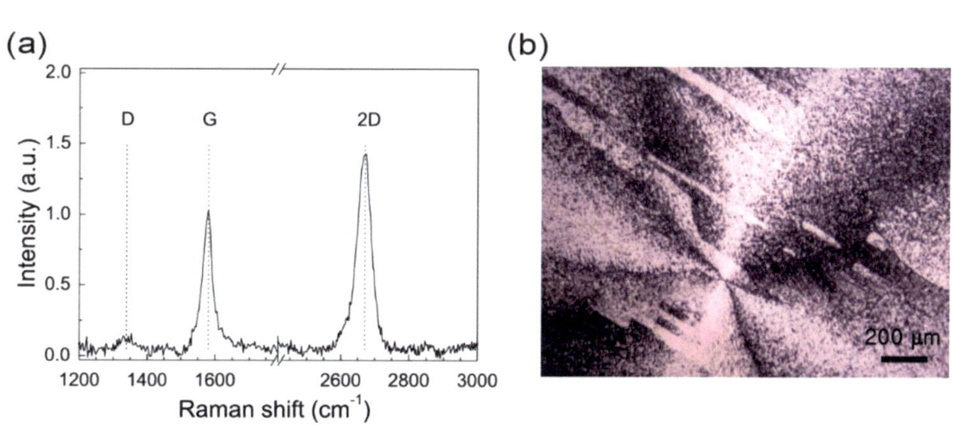

Figure 4.
(a) Raman spectra of the SLG transferred from the Cu foil onto the VOLET substrate. (b) Polarized optical microscope image of a spin-coated layer of commercial nematic liquid crystals on the SLG transferred to the VOLET substrate.

Work function [eV]	Dirac point energy [eV]	Hole mobility [$cm^2V^{-1}s^{-1}$]	Sheet resistance, [kΩ $square^{-1}$]
5.21	4.89	410	1.20

Table 1.
Summary of the electronic properties of the $FeCl_3$-doped SLG used.

Next, the transport characteristics of the SLG used were observed by assessing a liquid-gated lateral FET with SLG channels, a Gr-FET, as shown in **Figure 3(d)**. The lateral FET substrate was prepared using the VOLET substrate or a heavily doped n-type Si wafer substrate (0.05-ohm cm) with a thermally grown SiO_2 layer (300-nm-thick) as the gate dielectric for the OTFT, together with a laterally patterned metal source and drain electrodes consisting of a Cr layer (5.5-nm-thick) and a Au layer (50-nm-thick) formed on the substrate via a vacuum deposition process with a mask. The channel length (L) and width (W) of the FET were 50 μm and 1600 μm, respectively (see inset in **Figure 3(d)**). Regarding the transport characteristics of the SLG studied, a liquid-gated lateral Gr-FET was prepared using an FET substrate with acetonitrile (ACN) with 100 mM of tetrabutylammonium hexafluorophosphate ($TBAPF_6$). The channel of the studied SLG of the Gr-FET was gated through the ACN electrolyte with an Ag/AgCl (3.5 M KCl) reference electrode by sweeping the gate voltages from −0.8 to 0 V and then to +0.8 V with a sweep rate of 30 mV s^{-1} at V_{DS} = 100 mV. In general, the liquid-gate Gr-FET has better transfer characteristics than the conventional back-gate Gr-FET because the liquid gate exhibits higher capacitance than the back gate [49, 50].

For the SLG used here, the Gr-FET showed a clear asymmetrical V-shaped I_{DS}-V_G curve with a charge-neutral gate voltage (or Dirac point, V_{Dirac}) of ~0.54 V/$V_{Ag/AgCl}$. This large positive value of V_{Dirac} clearly indicates that the SLG used is p-type (hole) doped graphene due to the chlorination of graphene by $FeCl_3$ [49]. According to the V_{Dirac} value of the SLG, the energy level of the Dirac point E_D, relative to the vacuum level, can be estimated with regard to the redox potential of a probe material of ferrocene via the following relationship [51]:

$$E_D = \left[-\left(e V_{G,Dirac} - E_{1/2}(Fc/Fc+)\right) - 4.8\right] \text{eV}. \tag{2}$$

Here, 4.8 eV is the absolute energy level of the ferrocene and ferrocenium (Fc/Fc+) redox couple below the vacuum energy level, and $E_{1/2(Fc/Fc+)}$ = 0.45 eV [51]. From Eq. (2), the V_{Dirac} value of ~0.54 V/$V_{Ag/AgCl}$ for the SLG gives a Dirac point energy E_D of approximately ~4.89 eV. Note that the E_D value of 4.89 eV is higher than that (~4.49 eV) of epitaxial monolayer graphene [52], also confirming the p-type doping of the SLG. From the transfer characteristics, the carrier (hole) mobility μ of the SLG was also estimated using the relationship [53] of

$$\mu = \left(L/W C_g V_{DS}\right)\left(\Delta I_{DS}/\Delta V_G\right), \tag{3}$$

where C_g is the top-gate capacitance of graphene (~1.9 μF cm^{-2}) [50]. The estimated hole mobility for the SLG was approximately ~410 $cm^2\ V^{-1}\ s^{-1}$.

For the SLG studied here, Raman spectroscopy was also carried out using a confocal Raman system with a laser source operating at 514.5 nm (~1 mW on sample surface). As shown in **Figure 4(a)**, the Raman spectra of the SLGs studied here have two strong characteristic peaks, a G band at around ~1580–1600 cm^{-1}, due to the E_{2g} vibration of sp^2-bonded carbon atoms, and a 2D band at around ~2644–2665 cm^{-1}, which is a second-order type of vibrational mode caused by the scattering of phonons at the zone boundary [54, 55]. It can be observed that there are very small disorder-induced D bands around ~1340–1350 cm^{-1}, indicating the sparse formation of sp^3 bonds due to the relatively few defects in the SLGs studied.

From the Raman peak intensities, it was found that the ratios of the integrated Raman intensities of the G band to the 2D band for the $FeCl_3$-doped SLG were in the approximate range of 1.7–1.8, indicating that the SLGs studied here are high-quality monolayer graphene [55]. Moreover, from the peak positions, it was found that while the G and 2D peaks of the intrinsic undoped SLG are positioned at ~1579 cm^{-1} and ~2669 cm^{-1}, respectively, the G and 2D peak positions of the SLG used are correspondingly upshifted to ~1585 cm^{-1} and ~2677 cm^{-1}. Through a comparison of these with other examples in an earlier report of the relationship between the G and 2D peak positions of graphenes [55], it was verified that the SLG used here is p-type doped SLG.

Subsequently, the densities of the defects, the distances between the defects, and the porosities of nano-defects for the SLG were estimated from the ratio of the Raman intensities of the G bands to the D bands, I_D/I_G, as shown in the Raman spectra above. The density of the defects (n_D) and the distances between the defects (L_D) for the SLG, as estimated from the carbon amorphization trajectory ($I_D/I_G \sim 0.117$) [56, 57], were $n_D \sim 3.0 \times 10^{10}$ cm^{-2} and $L_D \sim 32.8$ nm, respectively, corresponding to a porosity of 9.4×10^{-2}%. This result clearly indicates that the SLGs studied here are nonporous high-quality SLGs with a negligible number of porous defects introduced during the synthesis and transfer steps.

Next, for this SLG, polarized optical microscopy was also carried out using SLG covered with commercial nematic liquid crystals (NLCs, Merck LC ZLI-2293) in a crossed polarization state [58]. As shown in **Figure 4(b)**, the polarized optical microscopic image of a spin-coated NLC layer on the SLG shows large graphene domains (with an average radius of the domains >100 μm) in the form of highly uniform optical retardation, in addition to small domains of several hundreds of nanometers in size [59, 60], clearly indicating that the SLG studied here is high-quality graphene with large-area graphene domains.

3. VOLETs with a doped CVD graphene source

3.1 Fabrication of SLG-based VOLETs

Figure 5 presents a schematic illustration of the structure used and the stages of the fabrication of the SLG-based VOLETs (Gr-VOLETs) with an ITO gate separated by an Al_2O_3 gate dielectric layer, a SLG source, organic channel layers, and an Al drain. The fabrication steps of the Gr-VOLET investigated are described below. To construct the Gr-VOLET, SLG (4 mm by 20 mm) was transferred onto a VOLET substrate, as mentioned above (Steps 1, 2). The source electrode used was $FeCl_3$-doped SLG. Next, organic semiconducting materials were deposited over the SLG source electrode regions; a channel layer of poly(para-phenylene vinylene) copolymer (known as SY, 70-nm-thick) was coated as an emissive channel layer by spin coating (Step 3), after which a 2-nm-thick electron injection layer of CsF and a 80-nm-thick drain electrode of Al were deposited on the top of the SY channel layer in sequence via thermal deposition at a rate of 0.05 nm s^{-1} under a base pressure of less than 2.7×10^{-4} Pa (Step 4). Finally, the fabricated device was encapsulated with an epoxy resin in a glove box. The photograph in **Figure 5** shows the microscopic morphology of the device cross section as observed by field emission scanning electron microscopy (SEM).

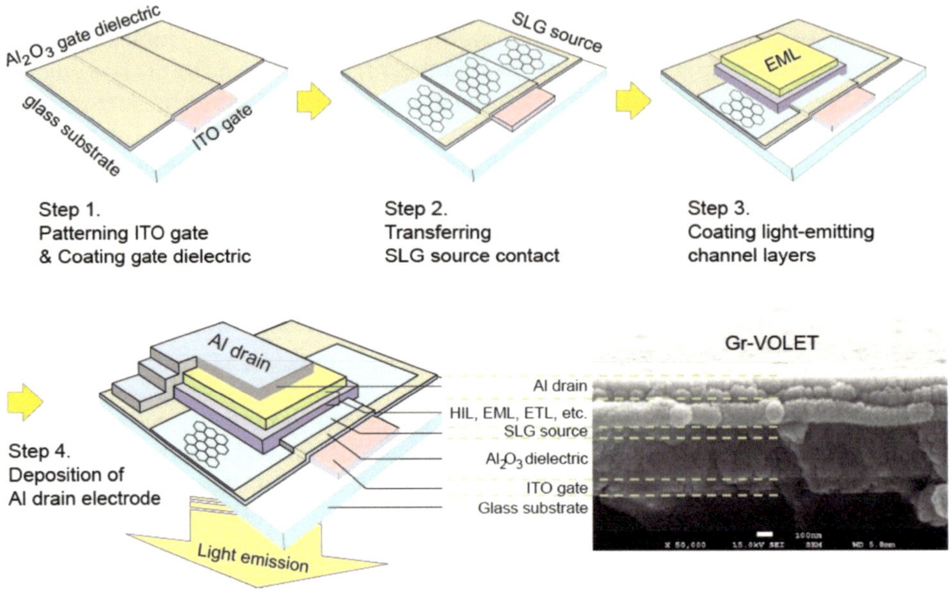

Figure 5.
Schematic illustration of the fabrication steps of a Gr-VOLET and a cross-sectional scanning electron microscopy (SEM) image of the Gr-VOLET with stacked layers of an ITO gate separated with an Al_2O_3 gate dielectric, a SLG source, organic channel layers, and an Al drain.

3.2 Operating characteristics of Gr-VOLETs

The operating characteristics of the Gr-VOLET were observed using a luminance meter in conjunction with two source meters. To operate the Gr-VOLETs, source-drain voltage V_{DS} (= $-V_{SD}$) on the Al drain and gate voltage, V_{GS} (or V_G), were applied with respect to the SLG source contact, held at the ground potential. During the operation of the Gr-VOLET, an electron injection occurs from the Al drain into the SY channel layer, and the hole injection from the SLG source to the SY channel layer can be modulated by controlling the gate voltage V_{GS} (or V_G), as discussed below.

Figure 6 shows the EL light emissions of a Gr-VOLET operating under different V_G values with a fixed V_{SD} of 3.8 V. As shown in **Figure 6(b)**, the EL light emission is uniform and bright (in the on-state) and evenly dark (off-state) over the entire surface of the active area for negative and positive V_G values, respectively. Hence, V_G essentially influences the radiative recombination process in the emissive channel layers.

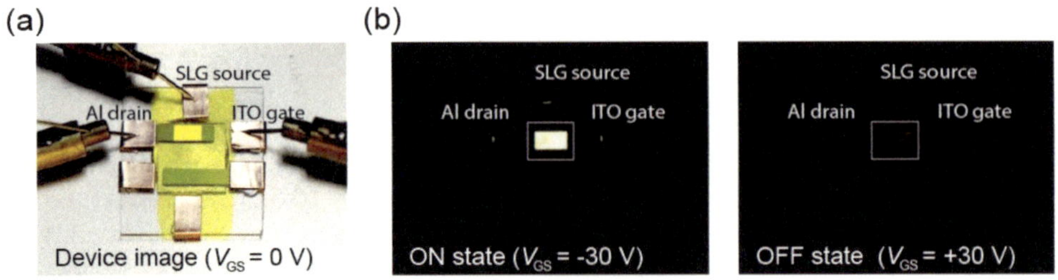

Figure 6.
EL light emission from a Gr-VOLET for different gate voltages, V_{GS}, for a given source-drain voltage, V_{SD}, of 3.8 V in bright (a) and dark (b) conditions (active area: 4 mm × 2 mm, white squares).

The output current and luminance characteristics of the Gr-VOLET were investigated as described below. For comparative purposes, the diode characteristics of the Gr-VOLET were also observed with the gate electrodes isolated from the external circuits (Gr-OLED). As shown in **Figure 7**, the current density-voltage (J_{SD}-V_{SD}) characteristics and luminance-voltage (L-V_{SD}) characteristics of the Gr-VOLET present the following key features: (i) the J_{SD}-V_{SD} characteristics are similar to the diode characteristics, as is generally observed in vertical organic FETs due to the short channel lengths [61]. (ii) Similar behaviors were observed in the L-V_{SD} characteristics; (iii) for a given V_{SD}, both J_{SD} and L increase with a decrease in V_G, even at low V_{SD} values, indicating that current modulation by V_G can change the EL emission brightness. Thus, the V_G dependent turn-on voltage, V_{onset}, can be reduced to well below V_{onset} of the Gr-OLED, and (iv) both J_{SD} and L also depend on the direction of change of V_G, that is, upward or downward, implying hysteretic behavior.

Interestingly, as shown in **Figure 7**, at $V_G = -40$ V, the Gr-VOLET with the doped SLG source exhibits high device performance, superior to that of the Gr-OLED, indicating an improved and balanced charge injection from the SLG source for negative V_G values. Conversely, at $V_G = +40$ V, J_{SD} and L of the Gr-VOLET are much lower, due to the switching off effect of the hole injections from the SLG source. The highest values of the peak on/off ratios for J_{SD} and L were ~10^2 and ~10^4, respectively, at $V_G = \pm 40$ V. Thus, the gate-bias-induced modulation effect of Gr-VOLET is shown to be quite efficient.

Next, the device performance, AR_{eff}, and PPC, of the Gr-VOLET were estimated in the on-state in comparison with a control OLED fabricated using an identical process on an ITO anode (ITO-OLED) (**Figure 8**). As shown in **Figure 8**, the Gr-VOLET exhibits EL luminance higher than that of the control OLED (ITO/SY/CsF/Al) in the $V_{SD} < 4.0$ V region. Moreover, the Gr-VOLET was shown to be more efficient than the control OLED (**Figure 8(c)**). For instance, at an EL luminance level of 500 cd m^{-2}, the Gr-VOLET emitted EL light with a current efficiency, η_C, of 7.13 cd A^{-1}, which is approximately 154% of the η_C result (4.64 cd A^{-1}) for the control ITO-OLED. Thus, the AR_{eff} value of the Gr-VOLET can be estimated to be 154% due to its full surface emission, just like the OLED. Thus, it is clear that the Gr-VOLET has a greatly enhanced AR_{eff} compared to other reported devices (**Table 2**). This result offers another substantial advantage: given the level of AR_{eff}, the brightness of the device can be maintained under a lower J_{SD}, providing a longer device lifetime

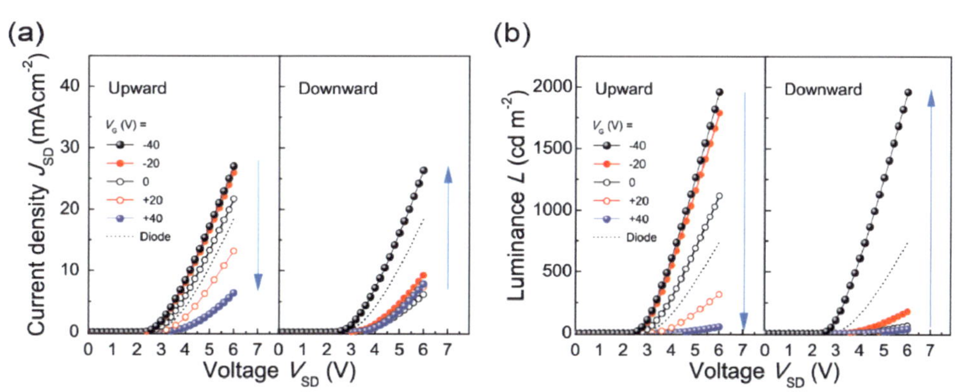

Figure 7.
Gate-voltage (V_G)-dependent current density-voltage (J_{SD}-V_{SD}) (a) and luminance-voltage (L-V_{SD}) (b) characteristics of a Gr-VOLET with a $FeCl_3$-doped SLG source for upward and downward changes in V_G. For comparison, the characteristics of a Gr-OLED (i.e., gate-disconnected Gr-VOLET) are also shown (dotted curves, OLED operations).

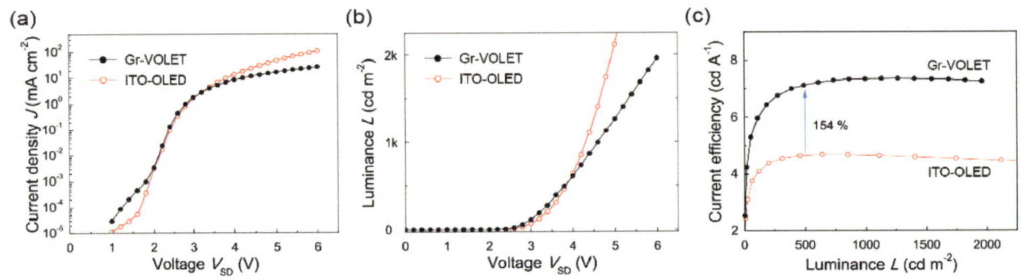

Figure 8.
J-V (a), L-V (b), and η_C-L (c) comparisons of the Gr-VOLET in the bright on-state (V_G = −40 V) with its ITO-based control OLED (ITO-OLED). Note that ITO-OLED = (ITO/SY/CsF/Al).

Devices	Reference	Source type	AR_{eff} [%]	*PPC* [%]
TFT + OLED	[15]			53
MIS-OLET	[19]			51
CNT-VOLET[a]	[32]	Porous CNT networks	98	6.2
Gr-VOLET	This work [49]	$FeCl_3$-doped SLG	154	5.2

[a] *Reference devices used the green phosphorescent emitter $Ir(ppy)_3$.*

Table 2.
Comparison of the effective aperture ratio, AR_{eff}, and parasitic power consumption, PPC, for various devices at a luminance level of 500 cd m^{-2}.

[62]. It is noteworthy that η_C (7.13 cd A^{-1}) for the Gr-VOLET was approximately 1.4 times higher than η_C (5.17 cd A^{-1}) of the ITO-OLED, possessing the optimized HIL of PEDOT:PSS. Thus, it is clear that the SLG source in the Gr-VOLET provides amplification of the emission and current efficiency, although further optimization of the electrodes is still possible.

Next, the *PPC* of the Gr-VOLET was deduced, which achieved luminance of 500 cd m^{-2} at V_{SD} = 3.82 V with an AR_{eff} rate of 154%, as discussed above. For the control ITO-OLED to emit a luminous flux through an aperture while transmitting 154% of its light, thereby matching the Gr-VOLET, it must emit 324 cd m^{-2}, requiring an applied voltage of 3.62 V. This indicates that 0.2 V of the V_{SD} (3.82 V) for the Gr-VOLET was dropped through its embedded transistor element, leading to a considerably reduced *PPC* of only 5.2%. This is much lower than that (6.2%) of the previous CNT-VOLET and the levels (>50%) for a TFT-OLED and a MIS-OLET (**Table 2**) [19, 32].

3.3 Charge injection process at SLG sources

At this point, our investigation turns to the hole injection mechanism at the interface between the SY channel layer and the SLG source. To be injected across the interface (SLG/SY), the holes must overcome the barrier height at the interface either via thermionic emission or tunneling processes [63–68]. **Figure 9(a)** shows examples of Fowler-Nordheim (F-N) curves [63–67] for the Gr-VOLET at various V_Gs during upward changes in V_G. All of the curves show two different hole injection processes with transition voltages (V_Ts), at which the injection mechanism changes to tunneling from Schottky thermionic emission [63–68]. In the figure, it is interesting to note that V_G affects both the Schottky thermionic emission and tunneling, indicating that V_T strongly depends on V_G. It is also noteworthy that because EL emission of the Gr-VOLETs occurs when V_{SD} > V_{onset} (V_{onset} > V_T), the main

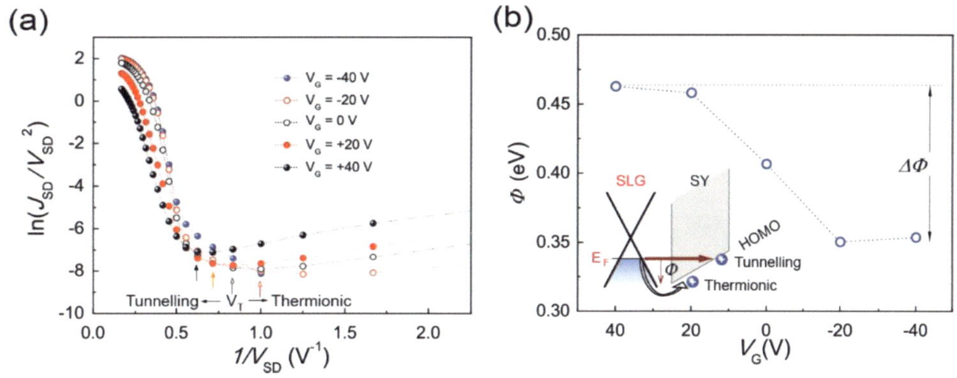

Figure 9.
(a) Fowler-Nordheim plots of the Gr-VOLET with the SLG source with various V_G values for upward V_G changes. (b) Gate-bias-modulated hole tunneling barrier height, Φ, extracted from the curve fittings in the hole-dominant regimes. The inset shows a schematic energy band diagram for the thermionic emission and tunneling at the interface between the SLG and the SY layer along the normal direction of the interface. Φ denotes the interface potential barrier height between the SLG and the SY channel layer.

hole injection process for EL emission is the tunneling injection in the Gr-VOLETs. According to the modified tunneling current model [70], the tunneling current density (J) of a single charge carrier through a triangular barrier at a polymer/metal junction is related to the potential barrier height (Φ) and the temperature (T),

$$\ln\left(J/V^2\right) = -P_1/V + \ln(P_2/V) - \ln\left[\sin(P_3/V)\right],$$
$$\text{with}\Phi = (2/3)\,\pi\,k_B T\,(P_1/P_3), \quad (4)$$

where k_B is the Boltzmann constant and P_i are parameters related to Φ. This relationship allows the F-N plots to be analyzed, and the potential barrier heights Φ at the SLG/SY interface between the Fermi level of SLG and the highest occupied molecular orbital (HOMO) level of the SY channel layer (~5.3 eV) [22] to be obtained (**Figure 9(b)**). As shown in the figure, the SLG/SY interface exhibited strong gate-bias-induced Φ modulation ($\Delta\Phi$) along the direction normal to the SLG/SY interface; that is, $\Delta\Phi$ at $V_G = \pm 40$ V was approximately 0.11 eV, leading to the effective modulation of the device performance of the Gr-VOLET tested here. Note that tunneling at the SLG/SY interface is the major process of hole injection, being responsible for the radiative recombinations of electron-hole pairs. Thus, the analysis of the tunneling process described above provides clear evidence that the Gr-VOLET operates based on vertical barrier height modulation along the direction normal to the SLG source surface (i.e., parallel to the direction of gate field), in contrast to the CNT-VOLET devices based on lateral modulation of Schottky barrier height along the horizontal direction on the CNT source surface (i.e., perpendicular to the direction of gate field) [31, 32], and differently from conventional graphene-based barristors that operate via modulation of the Schottky thermionic injection [69, 70].

The observations above show the working principle of the Gr-VOLET, as illustrated in the energy-level diagrams in **Figure 10**. At a given V_{SD}, a positive gate voltage induces an upward shift of the SLG Fermi level in a direction that increases the Φ, resulting in reduced hole tunneling injections into the HOMO level of the channel layer (SY). In contrast, a negative gate voltage induces a downward shift of the SLG Fermi level, decreasing Φ significantly (enhancing tunneling) and hence allowing increased hole injections and improved EL performance. Thus, together with the band-bending effect [31], the main operating mechanism of the Gr-VOLET

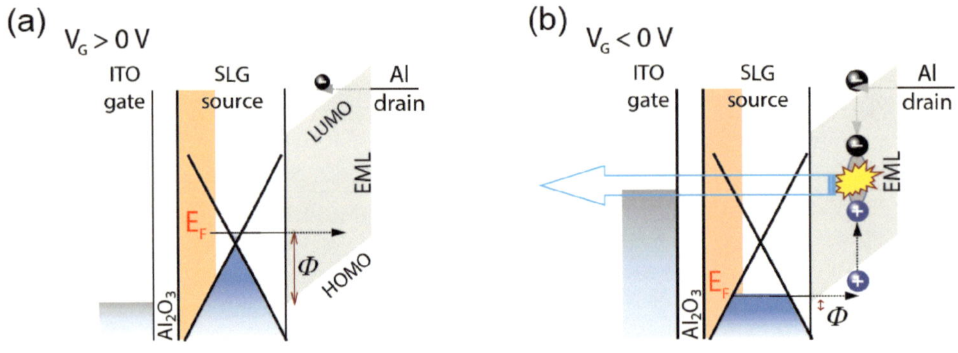

Figure 10.
Energy-level diagrams of the Gr-VOLET for high (a) and low Φs (b) at two distinct values of V_G and a given V_{SD}. Φ depicts the tunneling barrier height for the hole injection. E_F: Fermi energy level of the SLG source used.

is energy band matching, with charge balance achieved even without a HIL through gate voltage-induced modulation of the hole carrier tunneling injection at the p-type doped SLG source with $FeCl_3$.

In addition, notable instances of hysteresis were clearly observed, as shown above. Thus, bistable-like switching operations of a Gr-VOLET can allow novel applications for simple and inexpensive driving schemes together with low power consumption. However, this hysteresis effect may become an issue when attempting to realize high-quality grayscale outcomes and should be carefully, therefore, controlled when preparing the dielectric layer.

3.4 Inkjet-printing arrays of Gr-VOLET micro-pixels

Next, we turn our attention to a micro-pixel fabrication process for the Gr-VOLET using the inkjet-printing technique, as commonly used in solution-processable OLEDs [39, 71, 72]. Here, the inkjet technique used is based on the deposition of a small solvent drop onto an insulator layer, which can be easily redissolved and preferentially redeposited at the edge of the sessile drop (the contact line of the solvent drop), resulting in the formation of a via-hole with the shape of a crater, that is inkjet-etching [39].

To investigate the in situ formation of micro Gr-VOLET pixels created by means of inkjet-etching, an insulating polymer of poly(4-vinylpyridine) (P4VP) was introduced as a via-forming material, as P4VP is a hydrophilic polymer that dissolves in dimethyl formamide (DMF), toluene, chloroform, in lower alcohols, and in aqueous mineralic acids [71]. To fabricate a via-hole forming layer, a solution of P4VP with isopropanol (IPA) was spin-coated on top of the light-emitting channel layer of SY pre-coated onto a Gr-VOLET substrate (VOLET substrate/SLG/SY/P4VP). For micro-patterning, an etching solvent of chloroform for P4VP was inkjet-printed on top of the SY/P4VP layers (**Figure 11(a)**). This inkjet-printed solvent drop of chloroform can dissolve the P4VP layer, and the capillary flow of the solvent pushes the dissolved P4VP from the center to the contact line of droplet due to the coffee ring effect [39, 72–74], resulting the formation of the via-hole through the P4VP layer. Thus, after the deposition of even a single solvent droplet (~150 pL per droplet) on a 30-nm-thick P4VP film, the P4VP polymers are removed from the printed position and completely etched, forming via-holes through the P4VP layer, of which the inner and outer diameters are ~90 μm and ~120 μm, respectively, and finally uncovering the surface of the underlying SY layer. These P4VP via-holes on the light-emitting SY layer act as micro-patterned pixel openings

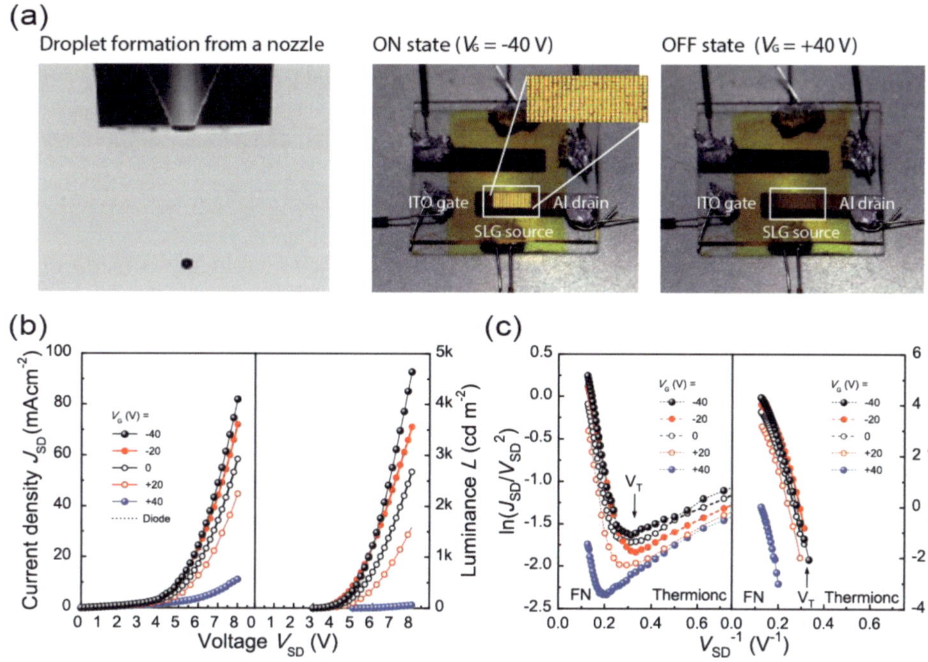

Figure 11.
(a) Left: A photographic image of a single drop of solvent ejected from an inkjet nozzle for drop formation. Right: Light emission from inkjet-printed Gr-VOLET pixels (white squares) for two different gate voltages V_Gs, at a fixed V_{SD} = 4.0 V. (b) Gate-voltage (V_G)-dependent current density-voltage (J_{SD}-V_{SD}) (left) and luminance-voltage (L-V_{SD}) (right) characteristics of the inkjet-printed Gr-VOLET pixels for various upward V_G changes. (c) Fowler-Nordheim plots, ln(J_{SD}/V_{SD}^2) vs. 1/V_{SD} (left) and ln(L/V_{SD}^2) vs. 1/V_{SD} (right) for the inkjet-printed Gr-VOLET pixels at various V_G values.

for the light-emitting active areas of the Gr-VOLETs. Then, to complete the device of an array of micro Gr-VOLETs, the CsF/Al/Ag cathode is deposited following the procedure described in Section 3.1.

Figure 11(a) also presents the switching behavior of EL light emissions from the array of micro Gr-VOLET pixels for two different gate voltages, V_Gs, at a fixed V_{SD} of 4.0 V. The photographic images show that only the isolated micro-patterned micro-pixel areas in contact with the SY layer of the Gr-VOLET pixels can emit EL light with a width of 85 μm, similar to the dimensions of the via-hole opening (90 μm). Note that the different solubility characteristics between the P4VP and SY polymers prevents any solvent erosion of the SY layer, as expected. As also clearly shown in the figure, the EL light emission from the micro Gr-VOLET pixels with a fixed droplet spacing of ~140 μm (~180 dpi) is highly bright (on-state) and dark (off-state) over the entire active area of the pixels for negative (−40 V) and positive values of V_G (+40 V), respectively. Hence, V_G clearly controls the radiative recombination process in the emissive pixel areas.

Next, the output characteristics of the inkjet-printed Gr-VOLET pixels were investigated. As shown in **Figure 11(b)**, the J_{SD}-V_{SD} and L-V_{SD} characteristics of the inkjet-printed Gr-VOLET pixels present features similar to those of the spin-coated Gr-VOLET (**Figure 7**). For example, at V_G = −40 V, the luminance reaches L ~ 1200 cd m^{-2} with a current efficiency of ~5.0 cd/A at V_{SD} = 6.0 V (V_{onset} = 3.0 V). In contrast, at V_G = +40 V, J_{SD} and L are greatly suppressed due to the switching off effect of the injection of holes from the SLG source, resulting in a peak on/off ratio of L of approximately 10^3 at V_G = ±40 V. **Figure 11(c)** presents F-N plots of the micro Gr-VOLET pixels for various V_Gs. As also shown in the figure, the negative slopes of the F-N plots clearly confirm that tunneling injection process is the major charge injection process for the light emission, being responsible for radiative recombinations of electron–hole pairs in the inkjet-printed Gr-VOLET pixels.

4. Conclusions

In summary, herein, graphene-based VOLETs have been explored, consisting of a nonporous, homogeneous, and p-doped SLG source with $FeCl_3$, an Al drain, and an emissive channel layer for efficient switching of the device performance using the gate voltage. Initially, we investigated transferred CVD SLG, which was used as the source electrode. It was found that the SLG used here was unintended p-doped SLG, exhibiting a Dirac point energy of ~4.9 eV and a work function of 5.2 eV with a shift of the Fermi level from the Dirac point and high hole mobility. It is shown that the high device performance capabilities of SLG-based VOLETs were mainly due to the p-doping effects, which were estimated quantitatively and analyzed based on the energy levels of the SLGs. It is also shown that low-drain-voltage operations and increased brightness with a high luminance on/off ratio ($\sim 10^4$) can be achieved even at high brightness for the Gr-VOLET without any HIL. Moreover, the current efficiency and effective aperture ratio of the Gr-VOLET were at least 150% higher than those of a control OLED, with low parasitic power consumption of 5%. These significant improvements of the device performance can be attributed to the gate-bias-induced modulation of the hole tunneling injection from the $FeCl_3$-doped SLG source into the emissive channel layer. Further, the feasibility of the simple fabrication process of micro Gr-VOLET pixels, that is, the inkjet-printing technique, was also proven.

The foregoing results demonstrate the notable device performance of the Gr-VOLET with graphene source, indicating considerable promise with respect to the development of high-performance VOLETs. The advances afforded by the Gr-VOLET with reliable switching performance, even at high luminance levels, clearly show its effective light-emitting transistor functionality and make it a feasible candidate for development of new voltage-driving light-emitting devices and/or highly integrated organic optoelectronics. Finally, it will be possible to apply advanced material layers to these Gr-VOLETs, which could lead to more efficient devices that operate even at low voltage levels, enabling the development of inexpensive, large-area, fast, and high-performance AM displays. Further improvements and characterizations are in progress and will be published elsewhere.

Acknowledgements

This work was supported by a grant from the National Research Foundation of Korea (NRF) funded by the Korean Government (MEST) (2017R1A2A1A17069729).

Author details

Byoungchoo Park*, Won Seok Lee, Seo Yeong Na, Jaewoo Park and In-Gon Bae
Department of Electrical and Biological Physics, Kwangwoon University, Seoul, Republic of Korea

*Address all correspondence to: bcpark@kw.ac.kr

References

[1] Tang CW, VanSlyke SA. Organic electroluminescent diodes. Applied Physics Letters. 1987;**51**:913-915. DOI: 10.1063/1.98799

[2] Burroughes JH, Bradley DDC, Brown AR, Marks RN, Mackay K, Friend RH, et al. Light-emitting diodes based on conjugated polymers. Nature. 1990;**347**:539-541. DOI: 10.1038/348352a0

[3] Yu G, Gao J, Hummelen JC, Wudl F, Heeger AJ. Polymer photovoltaic cells: Enhanced efficiencies via a network of internal donor-acceptor heterojunctions. Science. 1995;**270**: 1789-1791. DOI: 10.1126/science

[4] Chen HY, Hou J, Zhang S, Liang Y, Yang G, Yang Y, et al. Polymer solar cells with enhanced open-circuit voltage and efficiency. Nature Photonics. 2009;**3**: 649-653. DOI: 10.1038/nphoton.2009.192

[5] Sirringhaus H, Brown PJ, Friend RH, Nielsen MM, Bechgaard K, Langeveld-Voss BMW, et al. Two-dimensional charge transport in self-organized, high-mobility conjugated polymers. Nature. 1999;**401**:685-688. DOI: 10.1038/44359

[6] Bucella SG, Luzio A, Gann E, Thomsen L, McNeill CR, Pace G, et al. Macroscopic and high-throughput printing of aligned nanostructured polymer semiconductors for MHz large-area electronics. Nature Communications. 2015;**6**:8394. DOI: 10.1038/ncomms9394

[7] Baldo MA, O'Brien DF, You Y, Shoustikov A, Sibley S, Thompson ME, et al. Highly efficient phosphorescent emission from organic electroluminescent devices. Nature. 1998;**395**:151-154. DOI: 10.1038/25954

[8] Wang Q, Ding J, Ma D, Cheng Y, Wang L, Jing X, et al. Harvesting excitons via two parallel channels for efficient white organic LEDs with nearly 100% internal quantum efficiency: Fabrication and emission-mechanism analysis. Advanced Functional Materials. 2009;**19**:84-95. DOI: 10.1002/adfm.200800918

[9] Gu G, Forrest SR. Design of flat-panel displays based on organic light-emitting devices. IEEE Journal of Selected Topics in Quantum Electronics. 1998;**4**:83-99. DOI: 10.1109/2944.669473

[10] Shinar J, Shinar R. Organic light-emitting devices (OLEDs) and OLED-based chemical and biological sensors: An overview. Journal of Physics D: Applied Physics. 2008;**41**:133001. DOI: 10.1088/0022-3727/41/13/133001

[11] Lin YY, Gundlach DI, Nelson SF, Jackson TN. Pentacene-based organic thin-film transistors. IEEE Transactions on Electron Devices. 1997;**44**:1325-1331. DOI: 10.1109/16.605476

[12] Brody TP, Luo FC, Szepesi ZP, Davies DHA. A 6 × 6-in 20-lpi electroluminescent display panel. IEEE Transactions on Electron Devices. 1975;**22**:739-748. DOI: 10.1109/T-ED.1975.18214

[13] Nam WJ, Kim CY, Lee JH, Park SG, Han MK. 44.3: High-aperture ratio AMOLED pixel design employing VDD line elimination for reducing OLED current density. SID Symposium Digest of Technical Papers. 2005;**36**:1456-1459. DOI: 10.1889/1.2036282

[14] Nag M, Obata K, Fukui Y, Myny K, Schols S, Vicca P, et al. 20.1: Flexible AMOLED display and gate-driver with self-aligned IGZO TFT on plastic foil. In: SID Symposium Digest of Technical Papers. Vol. 45. 2014. pp. 248-251. DOI: 10.1002/j.2168-0159.2014.tb00068.x

[15] Ohta S, Chuman T, Miyaguchi S, Satoh H, Tanabe T, Okuda Y, et al.

Active matrix driving organic light-emitting diode panel using organic thin-film transistors. Japanese Journal of Applied Physics. 2005;**44**:3678-3681. DOI: 10.1143/jjap.44.3678

[16] Perumal A, Fröbel M, Gorantla S, Gemming T, Lüssem B, Eckert J, et al. Novel approach for alternating current (AC)-driven organic light-emitting devices. Advanced Functional Materials. 2012;**22**:210-217. DOI: 10.1002/adfm.201100747

[17] Kudo K, Tanaka S, Iizuka M, Nakamura M. Fabrication and device characterization of organic light emitting transistors. Thin Solid Films. 2003;**438**:330-333. DOI: 10.1016/s0040-6090(03)00751-x

[18] Park B, Takezoe H. Enhanced luminescence in top-gate-type organic light-emitting transistors. Applied Physics Letters. 2004;**85**:1280-1282. DOI: 10.1063/1.1784044

[19] Nakamura K, Hata T, Yoshizawa A, Obata K, Endo H, Kudo K. Improvement of metal–insulator–semiconductor-type organic light-emitting transistors. Japanese Journal of Applied Physics. 2008;**47**:1889-1893. DOI: 10.1143/jjap.47.1889

[20] Santato C, Capelli R, Loi MA, Murgia M, Cicoira F, Roy VAL, et al. Tetracene-based organic light-emitting transistors: Optoelectronic properties and electron injection mechanism. Synthetic Metals. 2004;**146**:329-334. DOI: 10.1016/j.synthmet.2004.08.028

[21] Muccini M. A bright future for organic field-effect transistors. Nature Materials. 2006;**5**:605-613. DOI: 10.1038/nmat1699

[22] Muhieddine K, Ullah M, Pal BN, Burn P, Namadas EB. All solution-processed, hybrid light emitting field-effect transistors. Advanced Materials. 2014;**26**:6410-6415. DOI: 1002/adma.201400938

[23] Capelli R, Toffanin S, Generali G, Usta H, Facchetti A, Muccini M. Organic light-emitting transistors with an efficiency that outperforms the equivalent light-emitting diodes. Nature Materials. 2010;**9**:496-503. DOI: 10.1038/nmat2751

[24] Ullah M, Armin A, Tandy K, Yambem SD, Burn PL, Meredith P, et al. Defining the light emitting area for displays in the unipolar regime of highly efficient light emitting transistors. Scientific Reports. 2015;**5**:8818. DOI: 10.1038/srep08818

[25] Muhieddine K, Ullah M, Maasoumi F, Burn PL, Namdas EB. Hybrid area-emitting transistors: Solution processable and with high aperture ratios. Advanced Materials. 2015;**27**:6677-6682. DOI: 10.1002/adma.201502554

[26] Hu Y, Lin J, Song L, Lu Q, Zhu W, Liu X. Vertical microcavity organic light-emitting field-effect transistors. Scientific Reports. 2016;**6**:23210. DOI: 10.1038/srep23210

[27] Zhang C, Chen P, Hu W. Organic light-emitting transistors: Materials, device configurations, and operations. Small. 2016;**12**:1252-1294. DOI: 10.1002/smll.201502546

[28] Liu CF, Liu X, Lai WY, Huang W. Organic light-emitting field-effect transistors: Device geometries and fabrication techniques. Advanced Materials. 2018;**30**:1802466. DOI: 10.1002/adma.201802466

[29] Muccini M, Toffanin S. Handbook of Organic Light-Emitting Transistors: Towards the Next Generation Display Technology. 1st ed. Hoboken, New Jersey: John Wiley & Sons, Inc.; 2016. 277 p. DOI: 10.1002/9781119189978

[30] Nakayama K, Pu YJ, Kido J. Surface-light-emitting transistors based on vertical-type metal-base organic transistors. Journal of the Society for Information Display. 2011;**19**:602-607. DOI: 10.1889/jsid19.9.602

[31] Liu B, McCarthy MA, Yoon Y, Kim DY, Wu Z. So F, et al. carbon-nanotube-enabled vertical field effect and light-emitting transistors. Advanced Materials. 2008;**20**:3605-3609. DOI: 10.1002/adma.200800601

[32] McCarthy MA, Liu B, Donoghue EP, Kravchenko I, Kim DY. So F, et al. low-voltage, low-power, organic light-emitting transistors for active matrix displays. Science. 2011;**332**:570-573. DOI: 10.1126/science.1203052

[33] Liu B, McCarthy MA, Iheanacho B, Wong WS, Rinzler AG. Recent developments of carbon nanotube enabled vertical organic light emitting transistors for OLED displays. SID Symposium Digest of Technical Papers. 2013;**44**:251-253. DOI: 10.1002/j.2168-0159.2013.tb06192.x

[34] McCarthy MA, Liu B, Cheney DJ, Lemaitre MG, Jayaraman R, Mativenga M, et al. QVGA AMOLED displays using the carbon nanotube enabled vertical organic light emitting transistor. SID Symposium Digest of Technical Papers. 2016;**47**:1796-1798. DOI: 10.1002/sdtp.11071

[35] Lee B, Chen Y, Cook A, Zakhidov A, Podzorov V. Stable doping of carbon nanotubes via molecular self-assembly. Journal of Applied Physics. 2014;**116**: 144503. DOI: 10.1063/1.4897550

[36] Geim AK, Novoselov KS. The rise of graphene. Nature Materials. 2007;**6**:183-191. DOI: 10.1038/nmat1849

[37] Ando T. The electronic properties of graphene and carbon nanotubes. NPG Asia Materials. 2009;**1**:17-21. DOI: 10.1038/asiamat.2009.1

[38] Bae S, Kim H, Lee Y, Xu X, Park JS, Zheng Y, et al. Roll-to-roll production of 30-inch graphene films for transparent electrodes. Nature Nanotechnology. 2010;**5**:574-578. DOI: 10.1038/nnano.2010.132

[39] Zhang Y, Liua C, Whalley DC. The impact of substrate temperature on the size and aspect ratio of inkjet-dissolved via holes in thin poly(4-vinyl phenol) dielectric layers. Applied Physics Letters. 2013;**102**:103303-103305. DOI: 10.1063/1.4795447

[40] Park B, Huh JN, Lee WS, Bae IG. Simple and rapid cleaning of graphenes with a 'bubble-free' electrochemical treatment. Journal of Materials Chemistry C. 2018;**6**:2234-2244. DOI: 10.1039/c7tc05695h

[41] Reina A, Jia X, Ho J, Nezich D, Son H, Bulovic V, et al. Large area, few-layer graphene films on arbitrary substrates by chemical vapor deposition. Nano Letters. 2009;**9**:30-35. DOI: 10.1021/nl801827v

[42] Li X, Cai W, An J, Kim S, Nah J, Yang D, et al. Large-area synthesis of high-quality and uniform graphene films on copper foils. Science. 2009;**324**:1312-1314. DOI: 10.1126/science.1171245

[43] Zhang Y, Zhang L, Zhou C. Review of chemical vapor deposition of Graphene and related applications. Accounts of Chemical Research. 2013;**46**:2329-2339. DOI: 10.1021/ar300203n

[44] Suk JW, Kitt A, Magnuson CW, Hao Y, Ahmed S, An J, et al. Transfer of CVD-grown monolayer graphene onto arbitrary substrates. ACS Nano. 2011;**5**:6916-6924. DOI: 10.1021/nn201207c

[45] Li X, Zhu Y, Cai W, Borysiak M, Han B, Chen D, et al. Transfer of large-area graphene films for high-performance transparent conductive

electrodes. Nano Letters. 2009;**9**:4359-4363. DOI: 10.1021/nl902623y

[46] Vinogradov NA, Simonov KA, Generalov AV, Vinogradov AS, Vyalikh DV, Laubschat C, et al. Controllable p-doping of graphene on Ir(1 1 1) by chlorination with $FeCl_3$. Journal of Physics Condensed Matters. 2012;**24**:314202. DOI: 10.1088/0953-8984/24/31/314202

[47] Yu YJ, Zhao Y, Ryu S, Brus LE, Kim KS, Kim P. Tuning the graphene work function by electric field effect. Nano Letters. 2009;**9**:3430-3434. DOI: 10.1021/nl901572a

[48] Panchal V, Pearce R, Yakimova R, Tzalenchuk A, Kazakova O. Standardization of surface potential measurements of graphene domains. Scientific Reports. 2013;**3**:2597. DOI: 10.1038/srep02597

[49] Park B, Lee WS, Na SY, Huh JN, Bae IG. Full-surface emission of graphene-based vertical-type organic light-emitting transistors with high on/off contrast ratios and enhanced efficiencies. Scientific Reports. 2019;**9**:6328. DOI: 10.1038/s41598-019-42800-y

[50] Ohno Y, Maehashi K, Yamashiro Y, Matsumoto K. Electrolyte-gated graphene field-effect transistors for detecting pH and protein adsorption. Nano Letters. 2009;**9**:3318-3322. DOI: 10.1021/nl901596m

[51] Padhy H, Huang JH, Sahu D, Patra D, Kekuda D, Chu CW, et al. Synthesis and applications of low-bandgap conjugated polymers containing phenothiazine donor and various benzodiazole acceptors for polymer solar cells. Journal of Polymer Science. Part A. Polymer Chemistry. 2010;**48**:4823-4834. DOI: 10.1002/pola.24273

[52] Gugel D, Niesner D, Eickhoff C, Wagner S, Weinelt M, Fauster T. Two-photon photoemission from image-potential states of epitaxial graphene. 2D Materials. 2015;**2**:045001. DOI: 10.1088/2053-1583/2/4/045001

[53] Schwierz F. Graphene transistors. Nature Nanotechnology. 2010;**5**:487-496. DOI: 10.1038/nnano.2010.89

[54] Ferrari AC. Raman spectroscopy of graphene and graphite: Disorder, electron–phonon coupling, doping and nonadiabatic effects. Solid State Communications. 2007;**143**:47-57. DOI: 10.1016/j.ssc.2007.03.052

[55] Wang QH, Jin Z, Kim KK, Hilmer AJ, Paulus GLC, Shih CJ, et al. Understanding and controlling the substrate effect on graphene electron-transfer chemistry via reactivity imprint lithography. Nature Chemistry. 2012;**4**:724-732. DOI: 10.1038/NCHEM.1421

[56] Cançado LG, Jorio A, Martins Ferreira EH, Stavale F, Achete CA, Capaz RB, et al. Quantifying defects in graphene via Raman spectroscopy at different excitation energies. Nano Letters. 2011;**11**:3190-3196. DOI: 10.1021/nl201432g

[57] Huang S, Dakhchoune M, Luo W, Oveisi E, He G, Rezaei M, et al. Single-layer graphene membranes by crack-free transfer for gas mixture separation. Nature Communications. 2018;**9**:2632-2642. DOI: 10.1038/s41467-018-04904-3

[58] Kim DW, Kim YH, Jeong HS, Jung HT. Direct visualization of large-area graphene domains and boundaries by optical birefringency. Nature Nanotechnology. 2012;7:29-34. DOI: 10.1038/nnano.2011.198

[59] Huang PY, Ruiz-Vargas CS, van der Zande AM, Whitney WS, Levendorf MP, Kevek JW, et al. Grains and grain boundaries in single-layer graphene atomic patchwork quilts. Nature. 2011;**469**:389-392. DOI: 10.1038/nature09718

[60] Nemes-Incze P, Yoo KJ, Tapasztó L, Dobrik G, Lábár J, Horváth ZE, et al. Revealing the grain structure of graphene grown by chemical vapor deposition. Applied Physics Letters. 2011;**99**:023104. DOI: 10.1063/1.3610941

[61] Li SH, Xu Z, Yang G, Ma L, Yang Y. Solution-processed poly(3-hexylthiophene) vertical organic transistor. Applied Physics Letters. 2008;**93**:213301. DOI: 10.1063/1.3030990

[62] Tsujioka T, Fujii H, Hamada Y, Takahashi H. Driving duty ratio dependence of lifetime of tris(8-hydroxy-quinolinate)aluminum-based organic light-emitting diodes. Japanese Journal of Applied Physics. 2001;**40**:2523-2526. DOI: 10.1143/jjap.40.2523

[63] Köhler A, Bässler H. Handbook of Electronic Processes in Organic Semiconductors: An Introduction. Weinheim, Germany: Wiley-VCH Verlag GmbH & Co. KGaA; 2015. pp. 424. DOI: 10.1002/9783527685172

[64] Fowler RH, Nordheim L. Electron emission in intense electric fields. Proceedings of the Royal Society A. 1928;**119**:173-181. DOI: 10.1098/rspa.1928.0091

[65] Sze SM. Handbook of Physics of Semiconductor Devices. 2nd ed. Hoboken, New Jersey: John Wiley & Sons, Inc.; 1981 868 pp

[66] Chiguvare Z, Parisi J, Dyakonov V. Current limiting mechanisms in indium-tin-oxide/poly3-hexylthiophene/aluminum thin film devices. Journal of Applied Physics. 2003;**94**:2440-2448. DOI: 10.1063/1.1588358

[67] Blum R, Sprave M, Sablotny J, Eich M. High-electric-field poling of nonlinear optical polymers. Journal of the Optical Society of America B. 1998;**15**:318-328. DOI: 10.1364/josab.15.000318

[68] Saker BK, Khondaker SI. Thermionic emission and tunneling at carbon nanotube-organic semiconductor interface. ACS Nano. 2012;**6**:4993-4999. DOI: 10.1021/nn300544v

[69] Yang H, Heo J, Park S, Song HJ, Seo DH, Byun KE, et al. Graphene Barristor, a triode device with a gate-controlled Schottky barrier. Science. 2012;**336**:1140-1143. DOI: 10.1126/science.1220527

[70] Bartolomeo AD. Graphene Schottky diodes: An experimental review of the rectifying graphene/semiconductor heterojunction. Physics Reports. 2016;**606**:1-68. DOI: 10.1016/j.physrep.2015.10.003

[71] Hsueh HY, Yao CT, Ho RM. Well-ordered nanohybrids and nanoporous materials from gyroid block copolymer templates. Chemical Society Reviews. 2015;**44**:1974-2018. DOI: 10.1039/C4CS00424H

[72] Chang SC, Liu J, Bharathan J, Yang Y, Onohara J, Kido J. Multicolor organic light-emitting diodes processed by hybrid inkjet printing. Advanced Materials. 1999;**11**:734-737. DOI: 10.1002/(SICI)1521-4095(199906)11:9<734::AID-ADMA734>3.0.CO;2-D

[73] Singh M, Haverinen HM, Dhagat P, Jabbour GE. Inkjet printing-process and its applications. Advanced Materials. 2010;**22**:673-685. DOI: 10.1002/adma.200901141

[74] Wang J, Song C, Zhong Z, Hu Z, Han S, Xu W, et al. In situ patterning of microgrooves via inkjet etching for a solution-processed OLED display. Journal of Materials Chemistry C. 2017;**5**:5005-5009. DOI: 10.1039/C7TC01330B

Chapter 7

Tetradentate Platinum(II) Emitters: Design Strategies, Photophysics, and OLED Applications

Huiyang Li, Tsz-Lung Lam, Liangliang Yan, Lei Dai, Byoungki Choi, Yong-Suk Cho, Yoonhyun Kwak and Chi-Ming Che

Abstract

This chapter provides an overview of tetradentate platinum(II) emitters as a promising class of metal-organic phosphorescent dopants for organic light-emitting diodes (OLEDs). Tetradentate platinum(II) emitters showing blue, green, and red light emissions, which are essential for full color displays as well as white light emission, are reviewed and discussed in the context of molecular design and photophysical and electroluminescent properties. Emphasis is placed on the molecular structures, the nature of emissive excited states [including ligand-centered (LC), intra-ligand charge transfer (ILCT), metal-to-ligand charge transfer (MLCT), and excimeric and oligomeric metal-metal-to-ligand charge transfer (MMLCT)], the intermolecular interactions impacting photophysical attributes (e.g., emission energies, quantum yields, and decay times), and OLED device performances.

Keywords: organic light-emitting diodes, metal complexes, platinum, phosphorescence, high efficiency, operational lifetime

1. Introduction

Organic light-emitting diodes (OLEDs) are solid-state devices based on organic films sandwiched between two electrodes that convert electricity to luminous energy. Since the pioneering work of electroluminescence (EL) in 1987 [1], OLEDs have been attracting considerable attention because of their light weight, low driving voltage, low power consumption, fast response speed, and high frame rate for displays, making them suitable for various applications [2–5], e.g., wearable devices, virtual reality (VR), smart homes and cities, and imaging and sensing applications. Currently, products with OLED displays are found in several fields, ranging from micro-displays to TV applications, and notably in smartphones and personal computers.

Figure 1 shows the typical structure of an OLED device, including a number of thin layers, which individually facilitate charge transfer or light emission [6–9]. During operation, when a suitable voltage is applied, electrons are injected into

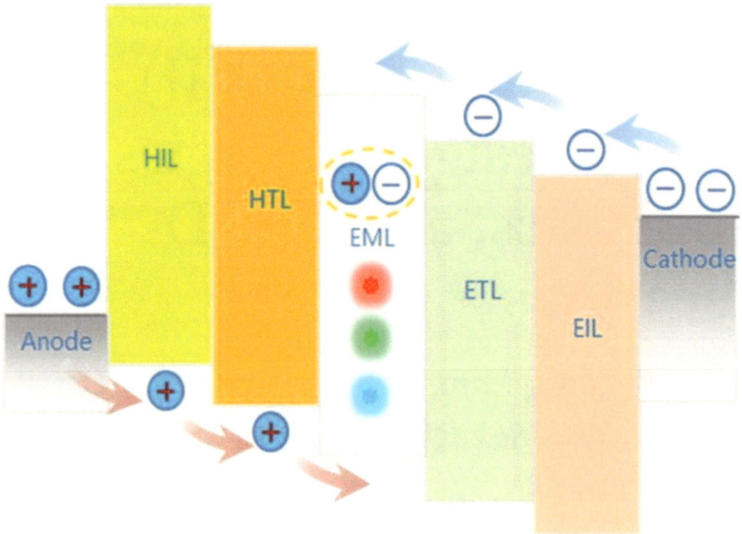

Figure 1.
Structure of an OLED device.

the electron transporting layer (ETL) from the metal cathode, which generally has a low work function (e.g., Al or Ag). To facilitate this process, a 0.5- to 1.0-nm thick electron injection layer (EIL) of LiF or CsF is usually deposited between the cathode and the ETL. Electrons migrate by hopping toward the anode. Meanwhile, holes are injected from the anode, which usually consists of a metal oxide mixture of SnO_2 (10%) and In_2O_3 (90%), namely, indium tin oxide (ITO). Following the anode, a hole injection layer (HIL) and a hole transporting layer (HTL) are typically required to promote hole transfer into the emission layer (EML), which includes the host matrix and dopant. Ideally, the recombination of electrons and holes takes place in the EML, subsequently populating the excited states that generate light emission. Obviously, the electron current must be well balanced with the hole current to avoid ohmic loss, which can be minimized by employing a hole blocking layer (HBL) between the ETL and EML and/or an electron blocking layer (EBL) between the HTL and EML. These blocking layers prevent the holes and electrons from leaving the EML without recombination. In addition, each layer requires materials with suitable highest occupied molecular orbital (HOMO) and lowest unoccupied molecular orbital (LUMO) energy levels. The development of auxiliary OLED materials (e.g., materials for the HTL, ETL, HBL, and EBL) has been the subject of previous reviews and is not discussed in this chapter [10–13].

As described above, the electrons and holes recombine and form neutral excitons in the EML. According to spin statistics, the recombination will populate singlet and triplet excited states in a ratio of 1:3, meaning that 75% of the electrically excited states are triplet states [14]. Pure organic molecules usually do not show spin-forbidden triplet emission (i.e., phosphorescence) at room temperature [15, 16], and normally only the singlet exciton is emissive, resulting in limited quantum efficiency, which presents a challenge for improving the efficiency of OLED devices. In 1998, the first successful applications of organometallic complexes as emissive dopant material in electroluminescent devices to generate phosphorescence were independently reported by Ma and Che [17] as well as Thompson and Forrest [18]. This revolutionary appraoch offers a viable means for the maximum use of electically generated excitons for electroluminescence and allowed a substantial leap forward in OLED performance. Since then, there has been increasing interest in the design and synthesis of new phosphorescent metal complexes, particularly those of Ru(II), Ir(III), and Pt(II), as OLED dopants [19–22].

2. Platinum(II) complexes as phosphorescent OLED dopants

Phosphorescent metal complexes remain mainstream OLED emitters, because relative to pure organic fluorophores, they can more efficiently harvest excitons for light emission. Phosphorescent metal complexes typically possess a transition metal ion with a high atomic number, e.g., Ru(II), Ir(III), or Pt(II), that can induce strong spin-orbit coupling (SOC), giving rise to ultrafast intersystem crossing (ISC) from the singlet to triplet states and promoting spin-forbidden triplet radiative decay. This triplet harvesting mechanism theoretically enables complete utilization of the excitons generated by electron-hole recombination for light emission, leading to a much higher efficiency and luminance. Reasonable phosphorescent metal OLED emitters should exhibit the following traits: (i) high phosphorescent quantum yields (i.e., >70%) when doped in a solid matrix, (ii) tunable emission color covering the blue, green, and red spectral regions (essential for full color displays), and (iii) superior thermal, chemical, and electrochemical stabilities for vacuum deposition and operation. The plethora of literature examples demonstrated that phosphors based on Ir(III) and Pt(II) can meet these requirements and generally outperform other metal phosphors [20–22]. In the past two decades, extensive research efforts on OLED emitters have been devoted to the development of phosphorescent Ir(III) and Pt(II) complexes and to investigations of their photophysical, electrochemical, and electroluminescent characteristics.

Phosphorescent Pt(II) complexes are noted for their desirable photophysical properties; they have a square planar coordination geometry, are coordinatively unsaturated, and exhibit diverse highly emissive excited states, including ligand-centered (LC), intra-ligand charge transfer (ILCT), ligand-to-ligand charge transfer (LLCT), metal-to-ligand charge transfer (MLCT), and excimeric and oligomeric metal-metal-to-ligand charge transfer (MMLCT) states [21, 22]. Tailoring the emission attributes (i.e., energy, quantum yield, lifetime, and radiative and non-radiative decay rate constants) to suit specific OLED applications can be achieved by the rational design of ligands, which allows regulation of (i) the energy levels of the metal d orbitals, the π and π^* orbitals of the ligands, and subsequently the composition of the frontier molecular orbitals, excited-state dynamics, and the nature of the emissive excited state and (ii) the intermolecular interactions that can contribute to emission from an aggregated state and/or emission quenching. As is the case for all phosphorescent OLED emitters, engineering the emissive excited state of Pt(II) emitters with high metal character to keep the emission lifetime short is important, since saturation of electroluminescence, severe efficiency roll-off at high luminance, and poor operational stability could otherwise result.

The planar coordination geometry renders platinum(II) complexes susceptible to self-assembly in ground and/or excited states through intermolecular ligand π-π and/or Pt-Pt interactions. This intrinsic property usually leads to a considerable redshift in the absorption and emission energies attributed to the generation of the low-energy emissive MMLCT excited states with enhanced radiative decay rate constants [21, 22], which could be harnessed to provide unique access to long-range ordered luminescent supramolecular structures, non-doped NIR OLEDs, and single-doped white OLEDs (WOLEDs). Nevertheless, this aggregation behavior could be unfavorable for applications in RGB OLED panels, especially when the Pt(II) emitters are doped at high concentrations, due to the possible occurrence of aggregate emission, triplet-triplet annihilation (TTA) and aggregation-caused quenching (ACQ). Therefore, the appropriate management of intermolecular interactions/aggregation by modulating the 3D morphology and electromagnetic properties of the complexes is crucial in the molecular design of Pt(II) emitters for specific OLED applications in order to achieve optimum device performances (i.e., high color purity, device efficiency, and long operational lifetime).

In the context of ligand architecture, the employment of tetradentate ligands in the design of platinum(II) emitters has clear advantages in terms of both chemical and thermal stabilities and luminescent efficiency compared to their bidentate and tridentate ligand counterparts. Tetradentate ligands provide a more stable scaffold for the coordination of platinum by offering the strong chelation effect, which could suppress ligand dissociation and demetalation. In addition, the rigid tetradentate ligand framework could largely restrict the excited-state metal-ligand structural distortion that in turn facilitates radiative deactivation of the emissive excited state, boosting the emission quantum yield of the Pt(II) emitter.

This chapter aims to provide an overview of mononuclear Pt(II) emitters containing tetradentate ligands reported in the literature. To keep the chapter to a reasonable size, we restrict our discussions to several selected classes of tetradentate platinum(II) emitters and apologize to the contributors to this field whose contributions are not mentioned herein.

3. Tetradentate platinum(II) emitters

3.1 Platinum(II) porphyrin complexes

The first photophysical studies of Pt(II) porphyrins were reported in the early 1970s and were triggered by the physicochemical relevance of metalloporphyrins such as chlorophyll, haem, and vitamin B12, which serve key biological functions [23]. This class of complexes is known for their high stability against heat, solvents, and extreme pH as a result of the strong coordination of Pt(II) ions in rigid porphyrin scaffolds. Pt(II) porphyrins exhibit intense absorptions in the visible region, and their electronic spectra are characterized by a Soret band at approximately 400 nm and two Q bands between 500 and 600 nm, which are attributed to porphyrin-centered $^1\pi$-π^* electronic transitions. The triplet formation yields for Pt(II) porphyrins were reported to be close to unity due to an ultrafast intersystem crossing process that occurs on a sub-ps time scale [24]. These complexes typically display strong saturated red to near infrared (NIR) phosphorescence, depending on the structures of macrocycles, with long decay times of tens of microseconds under anaerobic conditions because the emissive excited state is essentially ^{3}LC ($^3\pi$, π^*) in nature localized in the porphyrin ligand. For this reason, the emission properties can be rationally tuned by modifying the porphyrin ligands.

In 1998, Thompson and Forest reported the first use of a Pt(II) porphyrin complex, **Pt-1**, in electrophosphorescent devices (**Figure 2**) [18], which generated saturated red EL emission at 650 nm with a peak internal quantum efficiency (IQE) of 23%. Since then, the development of Pt(II) porphyrin complexes as phosphorescent emitters has attracted considerable interest [25–27].

Che et al. found that with the introduction of electron-deficient pentafluorophenyl rings at the meso positions of the porphyrin scaffold, **Pt-2** exhibited superior stability against oxidative degradation relative to that of the parental complex [Pt(TPP)] (TPP = 5,10,15,20-tetraphenylporphyrinato) (**Figure 2**) [28]. Saturated red (λ_{max} = 647 nm) porphyrin-centered phosphorescence with a long lifetime (τ_{em} = 60 μs) was observed for **Pt-2** in CH_2Cl_2 at room temperature. Saturated red OLED devices with different doping concentrations of **Pt-2** have

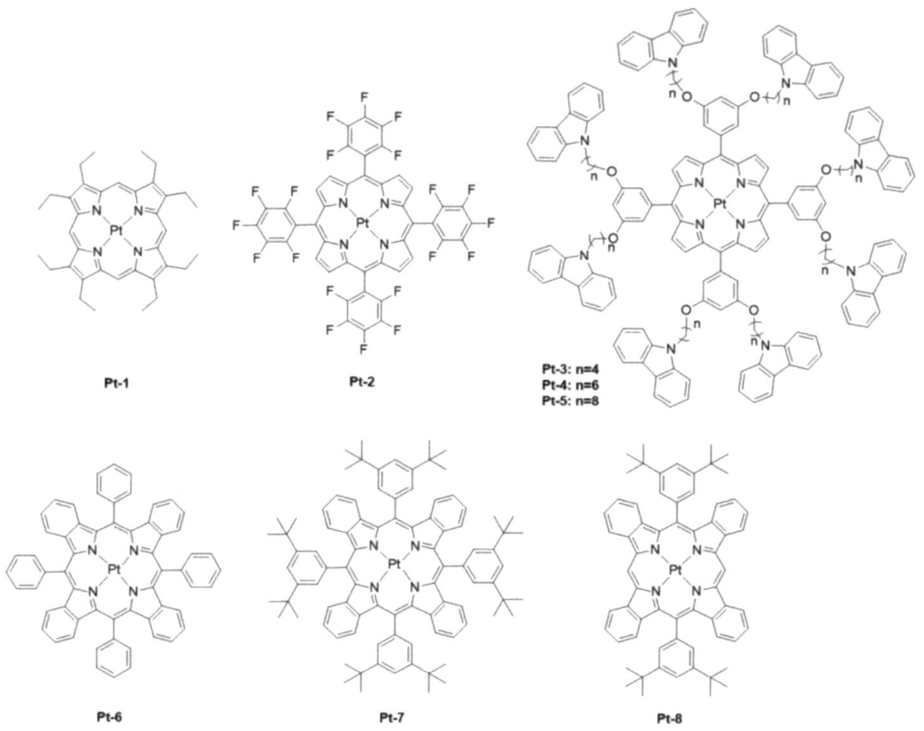

Figure 2.
*Chemical structures of platinum(II) porphyrin complexes **Pt-1–8**.*

been characterized. The EL spectra showed emission maxima at 655 nm, similar to that of the photoluminescent (PL) spectrum. At low driving voltages, emission from the host was not detected, indicating nearly complete energy transfer from the host. When a high voltage was applied, a blue emission from the host was observed, and its intensity increased with the driving voltage, which could be attributed to the saturation effect due to the long emission decay time of **Pt-2**. The efficiency increased until the dopant ratio reached 8% but decreased when the ratio was increased beyond 8%, which was rationalized by intermolecular quenching processes due to the high density of triplet excitons.

Wang and co-workers designed a group of platinum(II) porphyrin dendrimers (**Pt-3–5**) [29] with different alkyl chain lengths to adjust the distance between the Pt porphyrin core (the emissive center) and side carbazole groups (hole- and energy-transfer fragments) (**Figure 2**) and systematically investigated intra- and intermolecular energy-transfer mechanisms. In solution, both fluorescence at 357 nm from the carbazole moieties and phosphorescence at 660 nm from the Pt(II) porphyrin core were observed. The carbazole emission disappeared when doped in a solid matrix, indicating an efficient energy-transfer process. Intramolecular energy-transfer was facilitated by reducing the distance between the emitting core and the side carbazole groups. Solution-processed EL devices based on **Pt-3** were fabricated with a structure of [ITO/poly(styrenesulfonate)-doped poly(3,4-ethyle nedioxythiophene)/**Pt-3**/2,9-dimethyl-4,7-diphenyl-1,10-phenanthroline (BCP)/tris-(8-hydroxyquinoline)aluminum (Alq_3)/Ca/Al] to evaluate the optoelectronic properties. Bright phosphorescence (600 cd/m^2) with emission maxima at 540, 590, 660, and 720 nm was observed. The strong vibronic emission peaked at 660 was attributed to phosphorescence from the Pt(II) porphyrin cores, while the minor high-energy emission bands at 540 and 590 nm were suggested to be derived from thermally populated triplet states ("hot bands") or singlet states generated by TTA.

The data from devices prepared by spin coating these dendrimers as the non-doped emissive layer are better than those of small molecule Pt(II) porphyrin analogs.

The emission of platinum(II) porphyrin complexes can be further shifted to the NIR region by extending the π conjugation of the porphyrin ligand [30]. Schanze and co-workers developed a series of NIR-emitting platinum(II) di- and tetra-substituted benzoporphyrin complexes, **Pt-6–8** (**Figure 2**), to investigate structure-property relationships and the link between photophysical parameters and OLED performances. The photophysical parameters are listed in **Table 1**. Although the di-substituted porphyrin complex gives a higher Φ_{PL} (0.49) and longer τ_{em} (53.0 μs) than the tetra-substituted complexes (Φ_{PL} = 0.33–0.35; τ_{em} = 29.9–32.0 μs) in solution, the same trend does not hold in solid matrices, in which the emission lifetimes of **Pt-6–8** are comparable (45.7–57.5 μs). The EQEs of all the devices (8.0–9.2%) were similar. Consequently, the authors concluded that (i) the major non-radiative decays are associated with out-of-plane distortion of the porphyrin ligands, (ii) the rigid polymer matrix could help suppress the enhanced non-radiative decay of the more distorted complexes (**Pt-6** and **Pt-7**) in solid matrices, and (iii) the performance of platinum(II) benzoporphyrin-based OLEDs was observed to strongly correlate with the emission lifetime in the solid matrix.

Pt(II) porphyrins as OLED emitters generally exhibit high thermal stability and outstanding performance in saturated red and NIR devices in terms of color purity and EL efficiencies attributed to their narrow emission band and high emission quantum yields. Nevertheless, the practical interest of this class of Pt(II) emitters is limited by the long emission decay times, which would result in substantial efficiency loss at higher luminance. It is hypothesized that careful device design and the use of appropriate auxiliary materials to mitigate TTA, e.g., by using a double host to broaden the recombination zone, could be a strategy for improving the practicality of these materials.

3.2 Platinum(II) complexes supported by dianionic N_2O_2 ligands

3.2.1 Ligand systems and photophysical properties

Using "one-metal-one-ligand" approach to construct a stable luminescent platinum material, Che and co-workers developed the first non-porphyrin tetradentate aromatic N_2O_2 chelates, **Pt-9** and **Pt-10** (**Figure 3**), in 2003 [31]. The photophysical data of **Pt-9-12** are summarized in **Table 2**. These aromatic diimine-based Pt(II) complexes exhibit intense absorption bands at λ < 375 nm, which are attributed to

Complex	Φ_{PL} toluene	τ_{em} (μs) toluene	τ_{em} (μs) film	λ_{max} EL (nm)	Max EQE (%)
Pt-1	0.42	80.5	91.0[a]	650	4.0
Pt-2	0.09[b]	60.0[b]	—	655	—[c]
Pt-6	0.35	29.9	45.7[d]	773	8.0 ± 0.5
Pt-7	0.33	32.0	49.8[d]	773	9.2 ± 0.6
Pt-8	0.49	53.0	57.5[d]	777	7.8 ± 0.5

[a]*In polystyrene.*
[b]*In CH_2Cl_2.*
[c]*Max PE = 0.90 lm W^{-1}.*
[d]*In PVK:PBD.*

Table 1.
*Photophysical and OLED performance data for **Pt-1**, **Pt-2**, and **Pt-6–8**.*

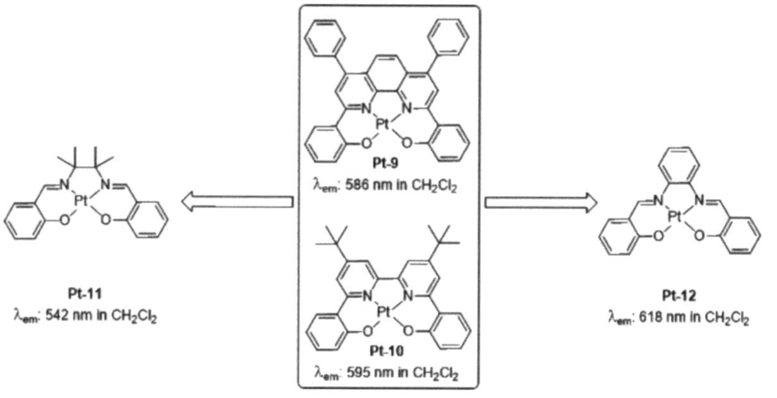

Figure 3.
Chemical structures of the platinum(II) complexes supported by dianionic N_2O_2 ligands, ***Pt-9–12****.*

		Emission		
Complex	**UV-Vis absorption in CH_2Cl_2, λ_{abs} (nm) (298 K) (ε, $\times10^4$ $mol^{-1}dm^3cm^{-1}$)**	**λ_{em} (nm)**	**τ_{em} (μs)**	**Φ_{PL}**
Pt-9	291 (3.92), 315 (3.40), 325 (3.23), 352 (2.58), 375 (2.47), 420 (0.52), 488 (sh, 0.67), 504 (0.72)	586	5.3	0.60
Pt-10	253 (4.10), 313 (1.84), 397 (0.840), 479 (0.294), 504 (sh, 0.252)	595	1.9	0.12
Pt-11	319 (1.31), 344 (1.67), 420 (0.58), 440 (0.54)	542	3.7	0.27
Pt-12	253 (4.16), 318 (2.49), 366 (3.61), 382 (3.41), 462 (0.93), 503 (sh, 0.86), 535 (0.99)	618	3.6	0.20

Table 2.
Photophysical data of ***Pt-9–12****.*

ligand-centered $^1\pi$-π^* transitions. The low-energy absorptions between 400 and 500 nm are assigned to the ^{1}ILCT transition (L → π^*, L = lone pair/phenoxide) mixed with ^{1}MLCT [dπ → π^*(diimine)] character. **Pt-9** and **Pt-10** display strong orange-red phosphorescence in CH_2Cl_2 at 298 K with λ_{max} values of 586 and 595 nm, respectively. The emission lifetimes and quantum yields are 5.3 μs and 0.6 for **Pt-9** and 1.9 μs and 0.1 for **Pt-10**. Owing to the ^{3}MLCT/^{3}ILCT nature of the emissive state, the emission lifetimes of **Pt-9** and **Pt-10** are significantly shorter than those of the aforementioned platinum(II) porphyrin complexes.

Schiff base ligands constitute another important class of N_2O_2 systems. The facile synthesis of Schiff base ligands, which can be prepared via one-pot multi-gram scale condensation reactions between substituted salicylic aldehydes and alkyl/aryl diamines, makes them an attractive ligand system for use in the synthesis of Pt(II) emitters. To elucidate structure-property relationships, Che and co-workers conducted a detailed investigation of a panel of Pt(II) Schiff base complexes with alkylene and arylene bridges (e.g., **Pt-11** and **Pt-12**; **Figure 3**) [32, 33]. The photophysical parameters are listed in **Table 2**. Similar to **Pt-9** and **Pt-10**, the absorption bands at $\lambda < 400$ nm are dominated by ligand-based $^1\pi$-π^* transitions, while those at $\lambda > 400$ nm are attributed to ^{1}MLCT and ^{1}ILCT [L → π^*] transitions. **Pt-11**, bearing a (tetramethyl)ethylene bridge, shows yellow-green emissions (λ_{em} = 541–546 nm) in solution with τ_{em} values of 3.4–3.9 μs and Φ_{PL} values of 0.18–0.27. When the nonconjugated bridge was replaced with a conjugated phenylene unit, as in **Pt-12**, a significant redshift in the emission λ_{max} to 608–628 nm in various solvents with τ_{em} values of 1.4–3.6 μs and Φ_{PL} of 0.10–0.26 was observed. In addition, the emission color can be finely tuned by attaching electron-donating or electron-withdrawing substituent(s) to the phenolate moieties of Schiff base ligands. The emission of these complexes displays moderate solvatochromic shift, and the emissive states were assigned to have mixed ^{3}ILCT [L → π*(diimine)] and ^{3}MLCT [d → π*(diimine)] characters. This assignment was further corroborated by the intermediate magnitude of their total zero-field spitting (ZFS) values between 14 and 28 cm^{-1}. These ZFS values lie between those of conventional Ru(II), Os(II), and Ir(III) ^{3}MLCT emitters (60–170 cm^{-1}) and those of Pd(II) and Rh(III) ^{3}IL emitters with ZFS < 1 cm^{-1}.

3.2.2 Chemical and thermal stability

Platinum(II) N_2O_2 complexes are generally stable in the solid state under ambient conditions. When dissolved in solution and exposed to light and air, **Pt-9** and **Pt-10** gradually decompose. By contrast, all Schiff base complexes are stable in common organic solvents such as EtOH, 2-propanol, DMSO, and CH_3CN under ambient conditions. All platinum(II) N_2O_2 complexes exhibit high thermal stability as assessed by thermal gravimetric analysis; **Pt-9** and **Pt-10** are stable up to 440 and 530°C, respectively. The decomposition temperatures of Pt(II) Schiff base complexes, including **Pt-11** and **Pt-12**, are in the range of 315–495°C. The introduction of $-CH_3$, *t*-Bu, or -F to the phenoxide moieties positively influenced the thermal stability of Pt(II) Schiff base complexes.

3.2.3 Electroluminescent properties

Devices with bis(2-(2-hydroxyphenyl)pyridine)beryllium ($Bepp_2$) as the host and **Pt-9** or **Pt-10** dopant as the emitting layer were fabricated: [ITO/N,N′-di(α-naphthyl)-N,N-diphenyl-(1,1-biphenyl)-4,4-diamine (NPB, 30 nm)/$Bepp_2$:**Pt-9** (or **Pt-10**) (30 nm)/LiF (0.5 nm)/Al (250 nm)]. All of the devices exhibited turn-on voltages ranging from 5 to 7 V, with yellow to yellow-green emissions. **Pt-10** showed a maximum luminance and power efficiency of 9330 cd m^{-2} (at 330 mA cm^{-2}) and 1.44 lm W^{-1} (at 40 mA cm^{-2}), respectively. Notably, although **Pt-9** exhibited a much higher emission quantum yield than **Pt-10** in solution, the EL performance of the former was inferior to that of the latter, which was attributed to the strong intermolecular quenching processes in **Pt-9**. Therefore, the bulky *t*-Bu groups in **Pt-10** are thought to play a vital role in suppressing intermolecular interactions.

The EL properties of platinum(II) Schiff base complexes were investigated. **Figure 4** shows two additional complexes, **Pt-13** and **Pt-14** [33, 34], discussed below, together with **Pt-11** and **Pt-12**. The EL spectra of devices with 4,4′-bis(carbazol-9-yl)biphenyl (CBP) as the host closely matched the corresponding PL spectra, suggesting that the EL originated from the same triplet excited states. The best device performances were obtained with dopant concentrations ranging from 1.5 to 4.5 wt%. At low dopant concentrations (<5.0 wt%), the devices exhibited yellow-green emission, and the efficiency was improved with increasing dopant concentrations. Additionally, the profile of the emission spectra remained unchanged. With doping concentration >5 wt%, the current efficiency (CE) was found to decrease, and the emission color changed due to the formation of excimers or aggregates. A maximum luminance of 9370 cd m^{-2} was achieved by optimizing the dopant concentration to 3 wt%. Notably, devices with simple structures, with $Bepp_2$ as the host and **Pt-13** as the dopant, can generate white emission, and the maximum luminance reached 3045 cd m^{-2}. Additionally,

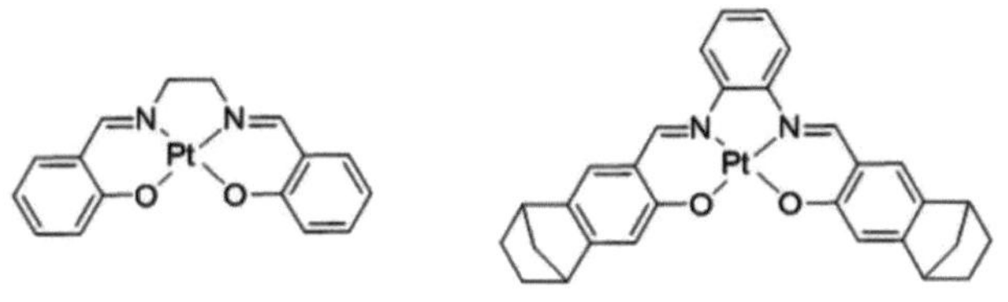

Figure 4.
*Chemical structures of platinum(II) Schiff base complexes **Pt-13** and **Pt-14**.*

the CIE coordinates of (0.33, 0.35) are close to those of white light (0.33, 0.33). Unlike **Pt-13**, no aggregate or excimer formation was observed for devices with 6.0 wt% of **Pt-11**, presumably due to the steric bulk of the (tetramethyl)ethylene bridges. Consequently, the performance of **Pt-11** was superior to that of **Pt-13**, with current and power efficiencies and luminance values up to 31 cd A^{-1}, 14 lm W^{-1}, and 23,000 cd m^{-2}, respectively, which are comparable to those of tris-cyclometalated iridium(III) complexes.

For red light-emitting materials, **Pt-12** achieved a current efficiency of 10.8 cd A^{-1} and an operational lifetime of >20,000 h at 100 cd m^{-2}. To suppress the intermolecular interactions as well as further optimize EL performance, a norbornene-based platinum(II) Schiff base complex, **Pt-14** (**Figure 4**) [34], was prepared. Sterically hindered norbornene moieties are highly effective in mitigating emission self-quenching. At a luminance of 1000 cd m^{-2}, **Pt-14** showed a current efficiency approximately 50% higher than that of **Pt-12** with the same device structure. In addition, the efficiency roll-off was reduced by 35%, benefiting from the lower self-quenching rate constant. By incorporating a wide bandgap iridium(III) complex as a co-dopant, high current and power efficiencies of 20.43 cd A^{-1} and 18.33 lm W^{-1}, respectively, were realized. In addition, the current efficiency could be maintained at 14.69 cd A^{-1} at a high luminance (1000 cd m^{-2}). More importantly, an operational lifetime of 18,000 h was realized at an initial luminance of 1000 cd m^{-2}, demonstrating that platinum(II) Schiff base complexes are promising red emitters for OLED displays.

Ease of synthesis, relatively short emission lifetime, high thermal stability, and decent emission quantum yield are traits that make platinum(II) N_2O_2 emitters attractive phosphorescent dopants, particularly for red OLEDs. Further research efforts in assessing and optimizing their operational stability in devices are anticipated.

3.3 Platinum(II) complexes supported by cyclometalated ligands

Incorporating anionic C-donor unit(s) into chromophoric ligands has been recognized as an effective strategy to enhance the luminescence of d^6 and d^8 transition metal complexes [35]. The same principle generally holds for tetradentate Pt(II) emitters. The tetradentate cyclometalated Pt(II) emitters reported in the literature typically feature high phosphorescence quantum yields of up to unity, which could be attributed to the following combined effects: (i) the rigid tetradentate ligand scaffold may help suppress excited-state structural distortion, thereby disfavoring non-radiative deactivation of the emissive excited state, (ii) the strongly σ-donating carbanion may destabilize the antibonding Pt $5d_{x2-y2}$ orbitals to a great extent, thus reducing the quenching of emissive states via the ^{3}d-d state, and (iii) the carbanion donor atom may also increase the metal character (e.g., ^{3}MLCT) and hence the radiative decay rate of the emissive excited states.

3.3.1 Pt(II) emitters with [O^N^C^N] ligands

In 2013, Che et al. developed the first phosphorescent platinum(II) complexes supported by tetradentate [O^N^C^N] ligands for white OLED and polymer organic light-emitting diode (PLED) applications [36]. These complexes were found to possess desirable physical properties as OLED emitters including high thermal stability with T_d > 400°C and ease of sublimation for vacuum deposition. Several follow-up studies on this family of Pt(II) emitters for high-efficiency OLEDs have been reported by the same group [22, 37–39].

3.3.1.1 Molecular design strategies

In general, for monochromic blue to yellow OLEDs, emission from monomeric Pt(II) complexes should be dominant, and aggregate emission should be minimized for achieving a high color purity (**Figure 5**). Early works showed that platinum(II) [O^N^C^N] complexes are prone to excimeric emission at elevated concentration. In attempts to address the excimer issues for realizing monchromic green and yellow OLEDs, Che and co-workers proposed a strategy to bolster the 3D configuration of the moelcular strucutre of the complexes to supress the intermolecular interactions via the introduction of rigid and bulky substituents, such as t-Bu groups and a norbornene moiety, to the ligand periphery, and the incorporation of bridging tertiary arylamine units or biphenyl groups with spiro linkages to the ligand frameworks. These modifications were found to effectively disfavor intermolecular interactions, evident by the diminished emission self-quenching and excimeric emissions in solution and in thin film at high concentrations. In addition, the corresponding devices showed improved device efficiencies and diminished efficiency roll-offs. Additionally, Pt(II) [O^N^C^N] emitters bearing a cross-shaped molecular structure (i.e., a spiro linkage) may also cause molecular entanglement in the amorphous state, which help prevent recrystallization of the emissive layer and prolong operational lifetimes.

Through deliberate molecular design and variations in the doping concentration, the extent of intermolecular interactions and aggregations of platinum(II) [O^N^C^N] emitters could be controlled and manipulated for red and NIR as well as white OLED applications based on aggregation and monomer/aggregation emissions, respectively (**Figure 5**). Instead of tuning the 3D configuration to limit intermolecular interactions, for these applications, adopting a relatively planar ligand scaffold and introducing fluorine substituent(s) at specific position(s),

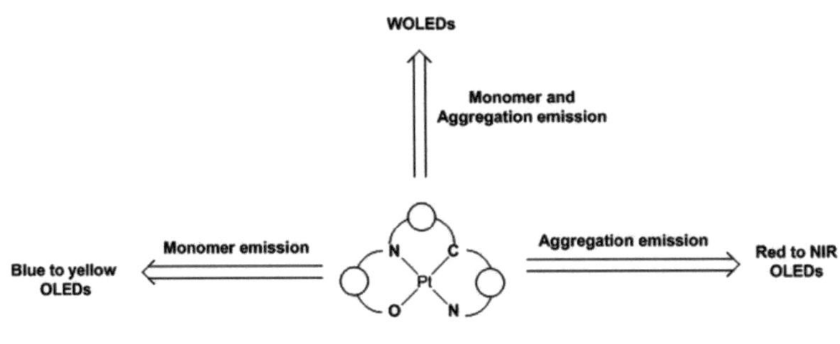

Figure 5.
Molecular design strategies for various OLED applications.

which favor intermolecular π-π and/or Pt-Pt interactions for low-energy aggregate emission from the excited states of dimers or oligomers, are preferred.

3.3.1.2 Photophysical properties and OLEDs based on monomer emission

The cyclometalated [O^N^C^N] ligand system is useful for the construction of robust and highly efficient Pt(II) emitters. Che et al. developed a panel of platinum(II) [O^N^C^N] emitters, i.e., **Pt-15–Pt-20** (**Figure 6**) [39], and systematically investigated their photophysical and electroluminescent properties, which are summarized in **Tables 3** and **4**.

The intense absorption bands of **Pt-15–20** at <300 nm are assigned to intraligand π-π* transitions, and the moderately intense bands at 430–450 nm with weak absorption at >460 nm are assigned to transitions with mixed MLCT and ILCT character. In degassed CH_2Cl_2 solutions, complexes **Pt-15–18** exhibit strong green to yellow luminescence (λ_{em} = 522–570 nm) with emission quantum yields and lifetimes in the range of 0.23–0.95 and 2.3–5.5 μs, respectively. With an additional t-Bu group at the *ortho*-position of the phenolate unit, **Pt-18** shows a much lower emission quantum yield (0.23) than **Pt-15–Pt-17** (0.77–0.95), indicating that the free rotation of this t-Bu group contributes to the increased non-radiative decay of the emissive state. Congeners **Pt-19** and **Pt-20**, with 6-5-6 metallacycles, are also highly efficient yellow (λ_{em} = 551 nm) and green (λ_{em} = 517 nm) phosphors, respectively. Their solution emission quantum yields (>0.80) and lifetimes (<5.1 μs) are similar to those of **Pt-15–17**. Femtosecond time-resolved fluorescence (fs-TRF) measurements suggested that **Pt-19** and **Pt-20** undergo an ultrafast intersystem crossing process with time constants of 0.44 and 0.15 ps, respectively. The emission origin for **Pt-15–20** was assigned to excited states with mixed ^{3}MLCT and ^{3}[L → π*] characters. Notably, **Pt-17–Pt-20** do not display excimer emissions in CH_2Cl_2 even at high concentrations (1.0×10^{-4} M), while **Pt-15** and **Pt-16** showed excimer emissions as a low-energy band (ca. 650 nm) under the same conditions. This finding suggests that the incorporation of a bridging tertiary amine or a spiro-fluorene linkage in the ligand framework would be as effective in suppressing the intermolecular

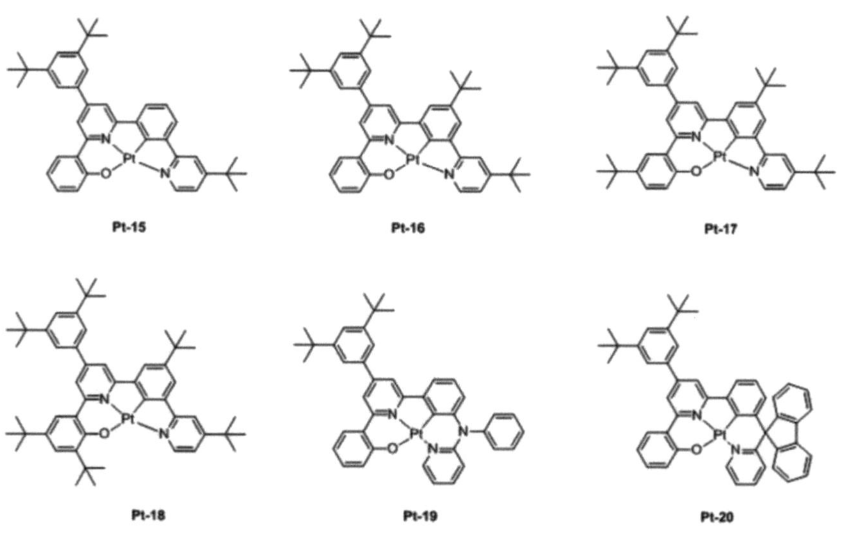

Figure 6.
Chemical structures of platinum(II) complexes ***Pt-15–20****.*

		Emission		
Complex	UV-Vis absorption in CH_2Cl_2, λ_{abs} (nm) (ε, $\times10^4$ $mol^{-1}dm^3cm^{-1}$)	λ_{em} (nm)	τ_{em} (μs)	Φ_{PL}
Pt-15	282 (4.5), 304 (sh, 3.3), 336 (sh, 1.8), 372 (1.9), 400 (sh, 1.1), 430 (sh, 0.8)	522	4.0	0.77
Pt-16	283 (4.4), 298 (sh, 3.7), 362 (sh, 1.6), 373 (1.7), 400 (sh, 1.0), 435 (sh, 0.69)	522	4.0	0.77
Pt-17	286 (4.4), 303 (sh, 3.2), 265 (sh, 1.5), 376 (1.8), 405 (sh, 0.98), 440 (sh, 0.75)	543	5.5	0.95
Pt-18	261 (5.1), 288 (5.4), 361 (sh, 1.5), 376 (2.2), 410 (sh, 1.2), 450 (sh, 0.85)	570	2.3	0.23
Pt-19	262 (4.4), 295 (sh, 3.5), 330 (2.2), 370 (sh, 1.1), 450 (sh, 0.27), 481 (sh, 0.21)	551	4.3	0.90
Pt-20	261 (sh, 5.0), 279 (5.4), 301 (sh, 3.6), 329 (1.8), 356 (1.7), 393 (0.72), 431 (sh, 0.38)	517	5.1	0.80

Table 3.
Photophysical data of ***Pt-15–Pt-20****.*

interactions/aggregation as introducing multiple bulky t-Bu substituents at the ligand periphery.

The EL properties of **Pt-15–Pt-18** were investigated in OLEDs based on the device structure [ITO/MoO_3 (5 nm)/di-[4-(N,N-ditolylamino)-phenyl]cyclohexane (TAPC, 50 nm)/4,4′,4″-tris(carbazole-9-yl)triphenylamine (TCTA):platinum complex (10 nm)/1,3,5-tri(m-pyrid-3-yl-phenyl) (TmPyPB, 50 nm)/LiF (1.2 nm)/Al (150 nm)], in which the emission layer was doped with **Pt-15–18** at different concentrations. The EL performance data are summarized in **Table 4**. At low doping concentrations, the EL spectra of all devices matched well with the corresponding solution-phase monomer emissions. With increasing dopant concentrations, aggregation emission was observed for **Pt-15–Pt-17**. Notably, aggregation emission was not observed for **Pt-18**, even at a high doping concentration of 15 wt%, revealing that the self-aggregation of **Pt-18** in EML is negligible. At an optimized doping concentration of 10 wt%, **Pt-18** achieved a maximum EQE of 27.1%, and this value dropped to 16.8% at a luminance of 10,000 cd m^{-2}.

The EL properties of **Pt-19** and **Pt-20** were studied in a device with the structure [ITO/MoO_3 (5 nm)/TAPC (50 nm)/TCTA:platinum(II) complex (10 nm)/TmPyPB or 2,4,6-tris(3-(3-(pyridin-3-yl)phenyl)phenyl)-1,3,5-triazine (Tm3PyBPZ, 50 nm)/LiF (1.2 nm)/Al (150 nm)] in which the doping concentration of the complexes ranged from 2 to 30 wt%. Similar to **Pt-15–Pt-18**, at a low doping concentration of 2 wt%, the emissions of both complexes are identical to the corresponding monomer emissions in solution. Increasing the doping concentration to 30 wt% caused a slight redshift for the **Pt-19**-based device, whereas only monomer emission was observed for **Pt-20** at the same doping level. This phenomenon could be rationalized by **Pt-19** to be more prone to undergo intermolecular interactions than those of **Pt-20**. In TmPyPB devices, a maximum EQE of 27.6%, CE of 104.2 cd A^{-1}, and PE of 109.4 lm W^{-1} have been achieved with 10 wt% **Pt-20**, and a maximum EQE of 26.0%, CE of 100.0 cd A^{-1}, and PE of 105.5 lm W^{-1} were achieved with **Pt-19** under the same conditions. At a high luminance (10,000 cd m^{-2}), the EQE of the devices based on **Pt-19** (30 wt%) or **Pt-20** (10 wt%) remained above 20%. To further optimize the PE of the OLEDs, TmPyPB was replaced with Tm3PyBPZ as the ETL. In Tm3PyBPZ devices, the driving voltage was significantly decreased. Consequently, the maximum PEs were improved to 118

	PE ($lm W^{-1}$)		CE ($cd A^{-1}$)		EQE (%)		
Complex (wt%)[a]	**Max.**	**At 10^4 $cd m^{-2}$**	**Max.**	**At 10^4 $cd m^{-2}$**	**Max.**	**At 10^4 $cd m^{-2}$**	**CIE (x, y)**
Pt-15 (2)	92.0	21.8	83.4	55.0	24.4	16.4	(0.32, 0.63)
Pt-15 (6)	25.2	5.2	25.5	16.4	13.5	8.6	(0.48, 0.50)
Pt-15 (12)	6.7	1.3	11.1	5.3	10.9	5.2	(0.60, 0.40)
Pt-16 (4)	61.5	10.5	68.3	31.4	19.7	8.6	(0.34, 0.62)
Pt-16 (8)	60.0	16.8	75.0	46.6	20.6	13.1	(0.35, 0.62)
Pt-16 (16)	48.6	16.1	51.0	43.4	20.4	17.1	(0.42, 0.56)
Pt-17 (2)	98.1	20.0	93.7	46.0	23.8	11.2	(0.39, 0.60)
Pt-17 (10)	94.3	32.1	90.0	68.1	24.8	18.8	(0.39, 0.60)
Pt-17 (15)	65.3	23.2	72.7	51.2	21.8	15.4	(0.40, 0.58)
Pt-18 (4)	82.1	7.4	86.1	22.8	22.7	6.6	(0.41, 0.57)
Pt-18 (10)	86.4	26.5	100.5	62.8	27.1	16.8	(0.41, 0.57)
Pt-18 (16)	91.0	32.6	94.0	73.8	26.3	19.1	(0.43, 0.56)
Pt-19 (2)	103.3	12.8	96.3	40.2	25.7	11.4	(0.42, 0.57)
Pt-19 (10)	105.5	20.6	100.0	55.4	26.0	14.4	(0.44, 0.55)
Pt-19 (16)	101.3	27.7	96.8	70.4	25.7	18.7	(0.45, 0.54)
Pt-19 (30)	80.7	27.8	82.5	68.4	24.8	20.5	(0.47, 0.52)
Pt-20 (2)	99.6	12.6	91.7	34.7	24.9	9.59	(0.29, 0.64)
Pt-20 (6)	106.7	31.1	101.1	73.1	26.9	19.1	(0.31, 0.64)
Pt-20 (10)	109.4	24.7	104.2	79.2	27.6	20.0	(0.31, 0.64)
Pt-20 (30)	95.7	28.4	90.0	66.9	24.0	17.9	(0.33, 0.63)
Pt-19 (10)[b]	118.0	22.5	94.3	48.0	25.3	12.5	(0.44, 0.55)
Pt-20 (10)[b]	126.0	24.4	98.8	52.0	26.4	13.6	(0.31, 0.63)

[a]*TmPyPB is used as the ETL.*
[b]*Tm3PyBPZ is used as the ETL.*

Table 4.
*OLED performance data for **Pt-15–Pt-20**.*

and 126 lm W^{-1} for the devices with **Pt-19** and **Pt-20** as dopants, respectively, and these values are comparable to those of the best iridium(III) OLED devices without out-coupling enhancement.

3.3.1.3 Aggregation-induced red and NIR OLEDs

The emission of Pt(II) complexes in aggregation forms is dramatically redshifted from that of monomers. Because of the increased metal character in the excited states (e.g., MMLCT) leading to the enhanced radiative decay rates, the emission lifetimes of aggregated Pt(II) emitters are usually short, in the range of 0.1–1 μs, which is fundamentally important for addressing the efficiency roll-off and the operational lifetime issues of phosphorescent OLEDs. In addition, this aggregation emission can be manipulated by tuning the doping concentration; this is particularly useful for the design of high-performance red and NIR OLEDs.

Recently, two series of platinum [O^N^C^N] complexes (**Figures 6** and 7), i.e., type-I (**Pt-21** and **Pt-22**) and type-II (**Pt-15, Pt-16** and **Pt-23**) [39, 40], which are prone to excited-state aggregation, were employed as emitting material in both doped and non-doped deep red and NIR devices; these complexes exhibited high EQE and low efficiency roll-off. For devices with neat complexes, high emission quantum yields were only realized with type-I complexes. For instance, when using a neat **Pt-21** film as the EML, the device demonstrated NIR emission with λ_{max} exceeding 700 nm and with an EQE of 15.84%. In addition, the EQE remained at 11.19% even at a high current density of 100 mA cm^{-2}. Of the doped devices, the device based on **Pt-16** (26 wt%) exhibited a deep red emission with λ_{max} of 661 nm, CIE coordinates of (0.63, 0.37), and an EQE value of 21.75% at a luminance of 1000 cd m^{-2}. The operational lifetimes at 90% initial luminance (LT90, L_0 = 100 cd m^{-2}) with 10 and 30 wt% **Pt-23** as the dopant were 59 and 374 h, respectively, demonstrating that aggregation-based devices would have longer lifetimes.

3.3.1.4 WOLEDs based on a single emitter

WOLED devices typically employ two or more co-dopants with different emission colors in the EML. Nevertheless, broad-band white light emission with a single Pt(II) emitter could be achieved when both the high-energy monomer emission and low-energy aggregation emission are harvested. In this case, a fine balance of the concentration of excited state monomers and excited state aggregation species is desired. Complex **Pt-22** displays both high-energy monomer emission at 482 nm and low-energy emission at 633–650 nm with emission quantum yields of up to 0.78 when doped into a solid matrix beyond 1.5 wt% [41]. This complex was first reported and used as a single emitter in white PLEDs by Che et al., and white

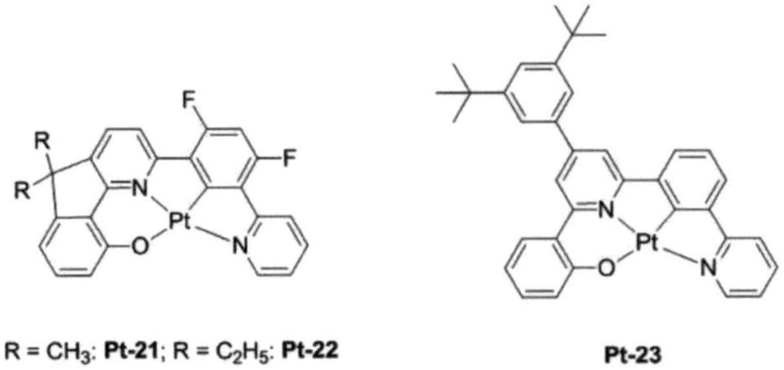

Figure 7.
*Chemical structures of platinum(II) complexes **Pt-21–Pt-23**.*

emission with an EQE of 11.51%, CIE coordinates of (0.41, 0.45), and a CRI of 74 at 1000 cd m^{-2} were realized with 16 wt% **Pt-22** as the dopant. To further optimize the performance of WOLEDs, devices with a structure of [ITO/MoO_3 (5 nm)/TAPC (50 nm)/host:7 wt% 7 (10 nm)/EML (50 nm)/LiF (1.2 nm)/Al (150 nm)] were fabricated. In these devices, 9-[3-[6-(3-carbazol-9-ylphenyl)pyridin-2-yl]phenyl] carbazole (26DCzppy) or TCTA/26DCzppy (1:1 in weight) was used as a single or double host, while TmPyPB or Tm3PyBPZ was used as the ETL. The EQE of the TmPyPB device dropped slightly with increasing luminance. To decrease the driving voltage, 26DCzppy was replaced with TCTA/26DCzppy (1:1 w/w), and the turn-on voltage was decreased from 3.5 to 3.0 V, resulting in an EQE of 23.2%. The turn-on voltage was further decreased to 2.7 V by replacing TmPyPB with Tm3PyBPZ. Consequently, the PE of the device reached a high value of 55.5 lm W^{-1}, which is comparable to the best reported values for a single emitter.

3.3.2 Pt(II) emitters with [N^C^C^N] ligands

In 2013, Li et al. developed two efficient blue-emitting tetradentate platinum complexes with a carbazolyl-pyridine motif integrated into the ligand scaffold. These complexes show emission quantum yields of up to 0.89, and the corresponding devices achieved excellent EQEs of up to 25%, highlighting the potential of these platinum emitters for blue OLED applications [42]. Subsequent works by the same group demonstrated that the carbazolyl-pyridine entity is also a versatile modular building block for various tetradentate dianionic cyclometalated N^C^C^N ligands, providing access to several new classes of efficient blue-, green-, and red-emitting platinum(II) complexes [43–46].

3.3.2.1 Molecular design strategies

The emission energies of the complexes in this family can be rationally and readily tuned by modifying the modular ligand scaffold, which consists of a cyclometalated chromophoric C^N unit and an auxiliary carbazolyl-pyridine group connected by a heteroatom or the heteroatom itself may be part of the chromophoric unit, as shown in **Figure 8**. Complex **Pt-24**, bearing a 4-phenylpyridine ring, shows red emission at 602 nm in solution at rt. Upon switching the 4-phenylpyridine group in the chromophoric unit to a pyrazole moiety to raise the LUMO energy, the emission of **Pt-25** is considerably blueshifted to 491 nm. The emission energy of the complexes can be further increased by reducing or breaking the π conjugation via manipulation of the chromophoric unit and/or the tethered group (**Pt-26–28**). For instance, by replacing carbazole with a 9,10-dihydroacridine group to interrupt the π conjugation, the emission maximum of **Pt-28** is blueshifted by 8 nm to 483 nm with respect to **Pt-25**.

This class of Pt[N^C^C^N] complexes was reported to be free from excimer-based emission, which was proposed to be a consequence of distortion of the molecular structure from planarity that disfavors intermolecular interactions [47]. Recently, Li and co-workers conducted a systematic photophysical study on derivatives of **Pt-25** and found that introducing substituents on the auxiliary unit dramatically influenced the emission spectral bandwidth and the nature of the emissive T_1 state through modulating the degree of mixing of ^{1}MLCT/^{3}MLCT characters with ^{3}IL state [48].

3.3.2.2 Red-emitting complexes and devices

Pt-24 is a representative red-emitting complex in this family [44]. This complex shows strong absorption bands at 250–400 nm ($\varepsilon = 2.4\text{–}6.4 \times 10^4$ cm^{-1} M^{-1}) attributable to $^1\pi$–π^* transitions localized on the cyclometalated tetradentate ligand.

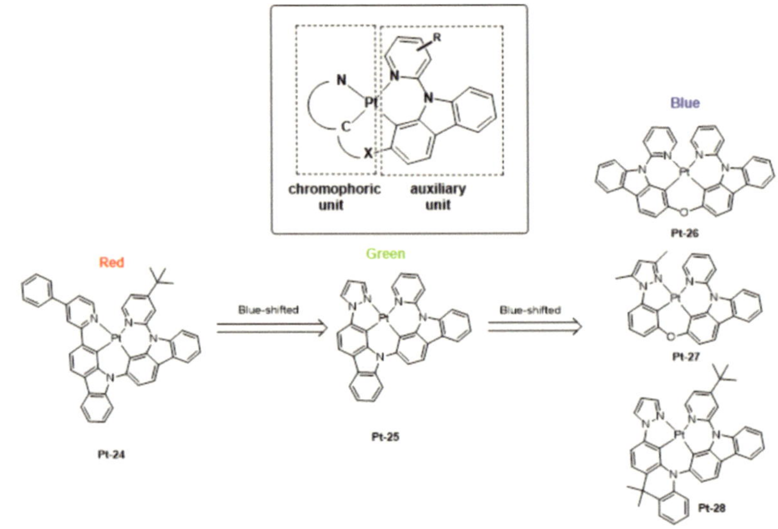

Figure 8.
Molecular design strategies for color tuning.

The moderately intense absorption, which can be assigned to the ^{1}MLCT transition appears at a longer wavelength of 450–550 nm (ε = 3900 cm^{-1} M^{-1}). Spin-forbidden triplet absorption is located beyond 560 nm (ε = 120 cm^{-1} M^{-1}) in CH_2Cl_2. **Pt-24** shows red emission at 602 nm in CH_2Cl_2 with an emission quantum yield of 34% at room temperature. A significant rigidochromic blueshift by ca. 30 nm to 574 nm is observed in the glassy solution (2-MeTHF) at 77 K, which is a sign of strong mixing of ^{1}MLCT/^{3}MLCT characters in the T_1 state.

The EL properties of **Pt-24** were studied with a device structure of [ITO/ dipyrazino[2,3-f:2′,3′-h]quinoxaline-2,3,6,7,10,11-hexacarbonitrile (HATCN, 10 nm)/NPB (40 nm)/10% **Pt-24**:CBP (25 nm)/BAlq (10 nm)/Alq_3 (40 nm)/LiF (1 nm)/Al (100 nm)], where BAlq is bis(2-methyl-8-quinolinolato)(biphenyl-4-olato)aluminum. The device showed an orange-red emission band at 606 nm, and this band was broader than that in solution. The EL spectrum also included a weak blue emission between 450 and 550 nm, originating from the hole transporting layer NPB. The device displayed a maximum EQE of 8.2% and an EQE of 7.8% at a luminance of 100 cd m^{-2}, which is not outstanding among the reported red-emitting metal complexes. However, the operational lifetime was encouraging. At an initial luminance of 1000 cd m^{-2}, the operational lifetime at 97% of the initial luminance (LT_{97}) was approximately 534 h, which is comparable to that of well-known iridium complexes with similar device structures, e.g., $Ir(ppy)_3$ and $(pq)_2Ir(acac)$. To remove the NPB emission as well as improve the efficiency, a 10-nm thick layer of 9,9′,9″-triphenyl-9H,9′H,9″H-3,3′:6′3″-tercarbazole (TrisPCz), with a higher LUMO level and triplet energy than the CBP host, was disposed between the HTL and EML. The maximum EQE was further increased to 11.8%, and the operational lifetime of LT_{97} was estimated to be 542 h at a luminance of 1000 cd m^{-2}. In addition, by replacing Alq_3 with 2,7-di(2,2′-bipyridin-5-yl)triphenylene (BPyTP), a device with the structure of [ITO/HATCN (10 nm)/NPB (40 nm)/TrisPCz (10 nm)/10% **Pt-24**:CBP (25 nm)/BAlq (10 nm)/BPyTP (40 nm)/LiF (1 nm)/ Al (100 nm)] was fabricated to decrease the driving voltage. Notably, the device showed a driving voltage of 3.6 V at a current density of 1 mA cm^{-2}, which was 1.6 V lower than that of the above devices with Alq_3 as the ETL. Importantly, the operational lifetime was also substantially improved, with an LT_{97} value of 638 h at a luminance of 1000 cd m^{-2}.

3.3.2.3 Green-emitting complexes and devices

Pt-25 displays strong green phosphorescence at 491 nm in CH_2Cl_2 at room temperature with emission quantum yields of 0.81 in CH_2Cl_2 and 0.90 in doped poly(methyl methacrylate (PMMA) films along with an exceptionally narrow spectral bandwidth with an full-width-at-half-maximum (FWHM) of 18 nm, which is comparable to those of quantum dots (25–40 nm) [46]. The authors attributed this phenomenon to localization of the T_1 state on the chromophoric unit.

To investigate the EL properties of **Pt-25**, OLEDs with the structure [ITO/PEDOT:PSS/NPB (30 nm)/TAPC (10 nm)/x% **Pt-25**:2,6-bis(N-carbazolyl)pyridine (26 mCPy, 25 nm)/2,8-bis(diphenylphosphoryl)-dibenzothiophene (PO15, 10 nm)/1,3-bis[3,5-di(pyridin-3-yl)phenyl] benzene (BmPyPB, 30 nm)/LiF (1 nm)/Al (90 nm)] were fabricated with dopant concentrations (x) ranging from 2 to 14%. The device with a doping concentration of 14% demonstrated a maximum EQE of 25.6%. Additionally, **Pt-25** was employed as the emitter in a device with a structure of [ITO/HATCN (10 nm)/NPB (40 nm)/x% **Pt-25**:CBP (25 nm)/BAlq (10 nm)/Alq_3 (40 nm)/LiF/Al] (x = 6, 10, and 20) to probe the operational stability. The device with a dopant concentration of 10% exhibited an operational lifetime of 70 h at 70% of the initial luminance (LT_{70}, L_0 = 2200 cd m^{-2}), corresponding to an LT_{70} of 32,000 h at an initial luminance of 100 cd m^{-2}. Additionally, in an optimized device structure of [ITO/HATCN (10 nm)/NPB (40 nm)/9-phenyl-3,6-bis(9-phenyl-9H-carbazol-3-yl)-9H-carbazole (TrisPCz; 10 nm)/10% **Pt-25**:3,3-di(9H-carbazol-9-yl)biphenyl (mCBP; 25 nm)/9,9′-(2,8-dibenzothiophenediyl)bis-9H-carbazole (mCBT; 8 nm)/BPyTP (40 nm)/LiF/Al], a maximum EQE of 22.1% and LT_{70} value of ca. 60,000 h were achieved at a luminance of 100 cd m^{-2}.

3.3.2.4 Blue-emitting complexes and devices

Breaking the π conjugation of ligand scaffolds can increase the T_1 energy for harvesting blue emission. By having all-six-membered chelate rings to interrupt the π conjugation, the O-bridged carbazolyl-pyridyl complex **Pt-26** shows a sky blue emission at 473 nm in a PMMA film with a high emission quantum yield of 0.83 and an emission lifetime of 3.8 μs [43]. A subtle disruption of π conjugation could also blueshift the emission. **Pt-28**, featuring a 9,10-dihydro-9,9-dimethylacridine subunit, displays a structured emission at 476 nm at 77 K [45], corresponding to CIE coordinates of (0.11, 0.30), which is 8 nm blueshifted from that of its carbazole analog, **Pt-25**. The **Pt-28**-doped PMMA film showed a high emission quantum yield of 0.68. Interestingly, the emission spectrum of **Pt-28** in CH_2Cl_2 at room temperature is dramatically broader than that of **Pt-25**, possibly due to the higher flexibility of the ligand.

Devices with the structure [ITO/HATCN (10 nm)/NPB (40 nm)/EBL/10% **Pt-28**:mCBP (25 nm)/HBL/BPyTP (40 nm)/LiF (1 nm)/Al (100 nm)] were fabricated and EL properties and operational lifetimes were examined. The EBL and HBL were arranged as follows: structure 1: no EBL/EML/BAlq (10 nm); structure 2: TrisPCz (10 nm)/EML/BAlq (10 nm); structure 3: no EBL/EML/mCBT (8 nm); and structure 4: TrisPCz (10 nm)/EML/mCBT (8 nm). The device with structure 1 exhibited a maximum EQE of 8.2% at a luminance of 1000 cd m^{-2}, and the LT_{70} was estimated to be 375 h at the same luminance, which corresponds to an LT_{70} value of 18,806 h at an initial luminance of 100 cd m^{-2}. Using TrisPCz to confine the electrons inside the EML, the device with structure 2 demonstrated a slightly improved peak EQE of 10.1% at 1000 cd m^{-2}, and the LT_{70} was estimated to be 416 h. Notably, when BAlq in structure 1 was replaced with a higher bandgap material (mCBT), the device with structure 3 displayed a peak EQE of 15.9%, and the LT_{70} was significantly prolonged to 635 h at 1000 cd m^{-2} and 31,806 h at 100 cd m^{-2}. Considering the advantages of TrisPCz (EBL) and mCBT (HTL) in the above devices (i.e., structures 2 and 3), these materials were employed in structure 4. As expected, this device achieved the best efficiency, namely, a peak EQE of 17.8%. However, the operational lifetime of LT_{70} decreased to 482 h at a luminance of 1000 cd m^{-2}.

Overall, tetradentate cyclometalated Pt(II) emitters have been demonstrated to exhibit high versatility in emission color tuning across RGB colors and white light, as well as superior photophysical and electroluminescent efficiencies and respectable operational lifetimes at practical luminance levels. While the performance metrics of this class of Pt(II) emitters are comparable to that of the best reported Ir(III) emitters in many aspects, more focused efforts should be directed at reducing the radiative lifetimes of these emitters by careful molecular design, which will be instrumental in further improving the operational stability of these complexes to meet the stringent standards required for commercialization.

4. Conclusions

Substantial room for innovation remains in OLED materials research, and the development of robust, high efficiency emitters for diverse applications remains a challenge both in academia and industry. While in the past decade, tris-(bidentate chelate) iridium(III) complexes have been seemingly edging out other classes of metal phosphors, it is remarkable that tetradentate platinum(II) emitters have demonstrated high performance and are being increasingly recognized by academia and industry as a competitive alternative. Importantly, the unique aggregation behavior and the associated photophysical properties afforded by their planar coordiantion geometry distinguish platinum(II) emitters from octahedral iridium(III) emitters. The unique photophysical properties of platinum(II) emitters render them well suited for some OLED applications using simple device structures such single-dopant WOLEDs and aggregation-based red and NIR OLEDs as covered in this review. In addition, appropriate molecular design of the ligand scaffold allows the regulation of the emissive excited states and the intermolecular interactions, which consequently offers flexibility in manipulating the emission characteristics of platinum(II) emitters to cater to various OLED applications. Indeed, sustained and concerted efforts between academia and industry have already realized successful application of tetradentate Pt(II) emitters in OLED devices in an industrial setting. It is without doubt that Pt(II) emitters, after full optimization, will meet the technical specifications including operational stability, required for commercialization. We hope the perspective described herein will spur interest among stakeholders and drive further development of tetradentate Pt(II) emitters for display and lighting applications.

Acknowledgements

This work was supported by the Major Program of Guangdong Basic and Applied Research (2019B030302009), Innovation and Technology Fund (ITS/224/17FP), Hong Kong Research Grants Council (HKU 17330416), the Basic Research Program of Shenzhen (JCYJ20170412140251576, JCYJ20170818141858021, and JCYJ20180508162429786), the National Key Basic Research Program of China (2013CB834802), Innovation and Technology Commission, Centre of Machine Learning for Energy Materials and Devices, a major initiative—Artificial Intelligence and Robotics cluster under InnoHK (AIR@InnoHK).

Conflict of interest

The authors declare no conflict of interest.

Author details

Huiyang Li[1], Tsz-Lung Lam[2,3], Liangliang Yan[1], Lei Dai[1], Byoungki Choi[4], Yong-Suk Cho[4], Yoonhyun Kwak[4] and Chi-Ming Che[2,3,5]*

1 Guangdong Aglaia Optoelectronic Materials Co., Ltd, Foshan, China

2 State Key Laboratory of Synthetic Chemistry, HKU-CAS Joint Laboratory on New Materials, Department of Chemistry, The University of Hong Kong, Hong Kong SAR, China

3 Centre of Machine Learning for Energy Materials and Devices, The University of Hong Kong, Hong Kong SAR, China

4 Samsung Electronics, Organic Material Lab, SAIT, Suwon-si, Gyeonggi-do, Korea

5 HKU Shenzhen Institute of Research and Innovation, Shenzhen, P. R. China

*Address all correspondence to: cmche@hku.hk

References

[1] Tang CW, Vanslyke SA. Organic electroluminescent diodes. Applied Physics Letters. 1987;**51**:913-915. DOI: 10.1063/1.98799

[2] Kalyani NT, Dhoble SJ. Organic light emitting diodes: Energy saving lighting technology—A review. Renewable and Sustainable Energy Reviews. 2012;**16**:2696-2723. DOI: 10.1016/j.rser.2012.02.021

[3] Tsutsui T, Takada N. Progress in emission efficiency of organic light-emitting diodes: Basic understanding and its technical application. Japanese Journal of Applied Physics. 2013;**52**:110001(1-9). DOI: 10.7567/jjap.52.110001

[4] Tiwari S, Singh M, Mishra SK, Shrivastava AK. Recent progress in organic light-emitting diodes. Journal of Nanoelectronics and Optoelectronics. 2019;**14**:1215-1224. DOI: 10.1166/jno.2019.2632

[5] Zampetti A, Minotto A, Cacialli F. Near-infrared (NIR) organic light-emitting diodes (OLEDs): Challenges and opportunities. Advanced Functional Materials. 2019;**29**:1807623(1-22). DOI: 10.1002/adfm.201807623

[6] Shuai Z, Peng Q. Excited states structure and processes: Understanding organic light-emitting diodes at the molecular level. Physics Reports-Review Section of Physics Letters. 2014;**537**:123-156. DOI: 10.1016/j.physrep.2013.12.002

[7] Thejokalyani N, Dhoble SJ. Novel approaches for energy efficient solid state lighting by RGB organic light emitting diodes—A review. Renewable & Sustainable Energy Reviews. 2014;**32**:448-467. DOI: 10.1016/j.rser.2014.01.013

[8] Wang J, Zhang F, Zhang J, Tang W, Tang A, Peng H, et al. Key issues and recent progress of high efficient organic light-emitting diodes. Journal of Photochemistry and Photobiology, C: Photochemistry Reviews. 2013;**17**:69-104. DOI: 10.1016/j.jphotochemrev.2013.08.001

[9] Chou PT, Chi Y. Phosphorescent dyes for organic light-emitting diodes. Chemistry - A European Journal. 2007;**13**:380-395. DOI: 10.1002/chem.200601272

[10] Tao Y, Yang C, Qin J. Organic host materials for phosphorescent organic light-emitting diodes. Chemical Society Reviews. 2011;**40**:2943-2970. DOI: 10.1039/c0cs00160k

[11] Shahnawaz, Swayamprabha SS, Nagar MR, RAK Y, Gull S, Dubey DK, et al. Hole-transporting materials for organic light-emitting diodes: An overview. Journal of Materials Chemistry C. 2019;**7**:7144-7158. DOI: 10.1039/c9tc01712g

[12] Kulkarni AP, Tonzola CJ, Babel A, Jenekhe SA. Electron transport materials for organic light-emitting diodes. Chemistry of Materials. 2004;**16**:4556-4573. DOI: 10.1021/cm049473l

[13] Huang Q, Cui J, Veinot JGC, Yan H, Marks TJ. Realization of high-efficiency/high-luminance small-molecule organic light-emitting diodes: Synergistic effects of siloxane anode functionalization/hole-injection layers, and hole/exciton-blocking/electron-transport layers. Applied Physics Letters. 2003;**82**:331-333. DOI: 10.1063/1.1536268

[14] Li Y. Organic Optoelectronic Materials. 1st ed. Cham/Heidelberg: Springer; 2015. p. 263. DOI: 10.1007/978-3-319-16862-3

[15] Leung MK, Chang CC, Wu MH, Chuang KH, Lee JH, Shieh SJ, et al. 6-N,N-diphenylaminobenzofuran-derived pyran containing fluorescent dyes: A new class of high-brightness red-light-emitting dopants for OLED. Organic Letters. 2006;**8**:2623-2626. DOI: 10.1021/ol060803c

[16] Swanson SA, Wallraff GM, Chen JP, Zhang W, Bozano LD, Carter KR, et al. Stable and efficient fluorescent red and green dyes for external and internal conversion of blue OLED emission. Chemistry of Materials. 2003;**15**:2305-2312. DOI: 10.1021/cm021056q

[17] Ma Y, Zhang H, Shen J, Che C. Electroluminescence from triplet metal-ligand charge-transfer excited state of transition metal complexes. Synthetic Metals. 1998;**94**:245-248. DOI: 10.1016/s0379-6779(97)04166-0

[18] Baldo MA, O'Brien DF, You Y, Shoustikov A, Sibley S, Thompson ME, et al. Highly efficient phosphorescent emission from organic electroluminescent devices. Nature. 1998;**395**:151-154. DOI: 10.1038/25954

[19] Chou PT, Chi Y. Osmium- and ruthenium-based phosphorescent materials: Design, photophysics, and utilization in OLED fabrication. European Journal of Inorganic Chemistry. 2006;**2006**:3319-3332. DOI: 10.1002/ejic.200600364

[20] Zysman-Colman E, editor. Iridium(III) in Optoelectronic and Photonics Applications. Chichester, West Sussex: John Wiley & Sons, Inc; 2017

[21] Kalinowski J, Fattori V, Cocchi M, Williams JAG. Light-emitting devices based on organometallic platinum complexes as emitters. Coordination Chemistry Reviews. 2011;**255**:2401-2425. DOI: 10.1016/j.ccr.2011.01.049

[22] Li K, So GMT, Wan Q, Cheng G, Tong WY, Ang WH, et al. Highly phosphorescent platinum(II) emitters: Photophysics, materials and biological applications. Chemical Science. 2016;**7**:1653-1673. DOI: 10.1039/c5sc03766b

[23] Eastwood D, Gouterman M. Porphyrins: XVIII. Luminescence of (Co), (Ni), Pd, Pt complexes. Journal of Molecular Spectroscopy. 1970;**35**:359-375. DOI: 10.1016/0022-2852(70)90179-7

[24] Atwater BW. Substituent effects on the excited-state properties of platinum *meso*-tetraphenylporphyrins. Journal of Fluorescence. 1992;**2**:237-246. DOI: 10.1007/BF00865282

[25] Ikai M, Ishikawa F, Aratani N, Osuka A, Kawabata S, Kajioka T, et al. Enhancement of external quantum efficiency of red phosphorescent organic light-emitting devices with facially encumbered and bulky Pt (II) porphyrin complexes. Advanced Functional Materials. 2006;**16**:515-519. DOI: 10.1002/adfm.200500492

[26] Sommer JR, Shelton AH, Parthasarathy A, Ghiviriga I, Reynolds JR, Schanze KS. Photophysical properties of near-infrared phosphorescent π-extended platinum porphyrins. Chemistry of Materials. 2011;**23**:5296-5304. DOI: 10.1021/cm202241e

[27] Retsek JL, Medforth CJ, Nurco DJ, Gentemann S, Chirvony VS, Smith KM, et al. Conformational and electronic effects of phenyl-ring fluorination on the photophysical properties of nonplanar dodecaarylporphyrins. The Journal of Physical Chemistry. B. 2001;**105**:6396-6411. DOI: 10.1021/jp004556k

[28] Che CM, Hou YJ, Chan MCW, Guo J, Liu Y, Wang Y. Meso-tetrakis(pentafluorophenyl)

porphyrinato platinum(II) as an efficient, oxidation-resistant red phosphor: Spectroscopic properties and applications in organic light-emitting diodes. Journal of Materials Chemistry. 2003;**13**:1362-1366. DOI: 10.1039/b212204a

[29] Li Y, Rizzo A, Salerno M, Mazzeo M, Huo C, Wang Y, et al. Multifunctional platinum porphyrin dendrimers as emitters in undoped phosphorescent based light emitting devices. Applied Physics Letters. 2006;**89**:061125(1-3). DOI: 10.1063/1.2335511

[30] Graham KR, Yang Y, Sommer JR, Shelton AH, Schanze KS, Xue J, et al. Extended conjugation platinum(II) porphyrins for use in near-infrared emitting organic light emitting diodes. Chemistry of Materials. 2011;**23**:5305-5312. DOI: 10.1021/cm202242x

[31] Lin YY, Chan SC, Chan MCW, Hou YJ, Zhu N, Che CM, et al. Structural, photophysical, and electrophosphorescent properties of platinum(II) complexes supported by tetradentate N_2O_2 chelates. Chemistry - A European Journal. 2003;**9**:1263-1272. DOI: 10.1002/chem.200390143

[32] Che CM, Chan SC, Xiang HF, Chan MCW, Liu Y, Wang Y. Tetradentate Schiff base platinum(II) complexes as new class of phosphorescent materials for high-efficiency and white-light electroluminescent devices. Chemical Communications. 2004;**40**:1484-1485. DOI: 10.1039/b402318h

[33] Che CM, Kwok CC, Lai SL, Rausch AF, Finkenzeller WJ, Zhu N, et al. Photophysical properties and OLED applications of phosphorescent platinum(II) Schiff base complexes. Chemistry - A European Journal. 2010;**16**:233-247. DOI: 10.1002/chem.200902183

[34] Zhou L, Kwong CL, Kwok CC, Cheng G, Zhang H, Che CM. Efficient red electroluminescent devices with sterically hindered phosphorescent platinum(II) Schiff base complexes and iridium complex codopant. Chemistry - An Asian Journal. 2014;**9**:2984-2994. DOI: 10.1002/asia.201402618

[35] Chi Y, Chou PT. Transition-metal phosphors with cyclometalating ligands: Fundamentals and applications. Chemical Society Reviews. 2010;**39**:638-655. DOI: 10.1039/b916237b

[36] Kui SCF, Chow PK, Tong GSM, Lai SL, Cheng G, Kwok CC, et al. Robust phosphorescent platinum(II) complexes containing tetradentate O^N^C^N ligands: Excimeric excited state and application in organic white-light-emitting diodes. Chemistry - A European Journal. 2013;**19**:69-73. DOI: 10.1002/chem.201203687

[37] Lai SL, Tong WY, Kui SCF, Chan MY, Kwok CC, Che CM. High efficiency white organic light-emitting devices incorporating yellow phosphorescent platinum(II) complex and composite blue host. Advanced Functional Materials. 2013;**23**:5168-5176. DOI: 10.1002/adfm.201300281

[38] Kui SCF, Chow PK, Cheng G, Kwok CC, Kwong CL, Low KH, et al. Robust phosphorescent platinum(II) complexes with tetradentate O^N^C^N ligands: High efficiency OLEDs with excellent efficiency stability. Chemical Communications. 2012;**49**:1497-1499. DOI: 10.1039/c2cc37862k

[39] Cheng G, Kui SCF, Ang WH, Ko MY, Chow PK, Kwong CL, et al. Structurally robust phosphorescent [Pt(O^N^C^N)] emitters for high performance organic light-emitting devices with power efficiency up to 126 lm w^{-1} and external quantum efficiency over 20%. Chemical Science. 2014;**5**:4819-4830. DOI: 10.1039/c4sc01105h

[40] Cheng G, Wan Q, Ang WH, Kwong CL, To WP, Chow PK, et al.

High-performance deep-red/near-infrared OLEDs with tetradentate [Pt(O^N^C^N)] emitters. Advanced Optical Materials. 2019;**7**:1801452(1-7). DOI: 10.1002/adom.201801452

[41] Cheng G, Chow PK, Kui SCF, Kowk CC, Che CM. High-efficiency polymer light-emitting devices with robust phosphorescent platinum(II) emitters containing tetradentate dianionic O^N^C^N ligands. Advanced Materials. 2013;**25**:6765-6770. DOI: 10.1002/adma.201302408

[42] Hang XC, Fleetham T, Turner E, Brooks J, Li J. Highly efficient blue-emitting cyclometalated platinum(II) complexes by judicious molecular design. Angewandte Chemie International Edition. 2013;**52**:6753-6756. DOI: 10.1002/anie.201302541

[43] Fleetham TB, Huang L, Klimes K, Brooks J, Li J. Tetradentate Pt(II) complexes with 6-membered chelate rings: A new route for stable and efficient blue OLEDs. Chemistry of Materials. 2016;**28**:3276-3282. DOI: 10.1021/acs.chemmater.5b04957

[44] Fleetham T, Li G, Li J. Efficient red-emitting platinum complex with long operational stability. ACS Applied Materials & Interfaces. 2015;**7**:16240-16246. DOI: 10.1021/acsami.5b01596

[45] Li G, Klimes K, Fleetham T, Zhu ZQ, Li J. Stable and efficient sky-blue organic light emitting diodes employing a tetradentate platinum complex. Applied Physics Letters. 2017;**110**:11301(1-5). DOI: 10.1063/1.4978674

[46] Li G, Fleetham T, Turner E, Hang XC, Li J. Highly efficient and stable narrow-band phosphorescent emitters for OLED applications. Advanced Optical Materials. 2015;**3**:390-397. DOI: 10.1002/adom.201400341

[47] Fleetham T, Li G, Li J. Phosphorescent Pt(II) and Pd(II) complexes for efficient, high-color-quality, and stable OLEDs. Advanced Materials. 2017;**29**:1601861. DOI: 10.1002/adma.201601861

[48] Li G, Wolfe A, Brooks J, Zhu ZQ, Li J. Modifying emission spectral bandwidth of phosphorescent platinum(II) complexes through synthetic control. Inorganic Chemistry. 2017;**56**:8244-8256. DOI: 10.1021/acs.inorgchem.7b00961